The
Council of
State Governments
CSG

T5-BQB-582

Division of Policy Analysis Services

2 | Economic Development in the States

The Changing Arena: State Strategic Economic Development

by Lee Walker, Policy Analyst
The Council of State Governments

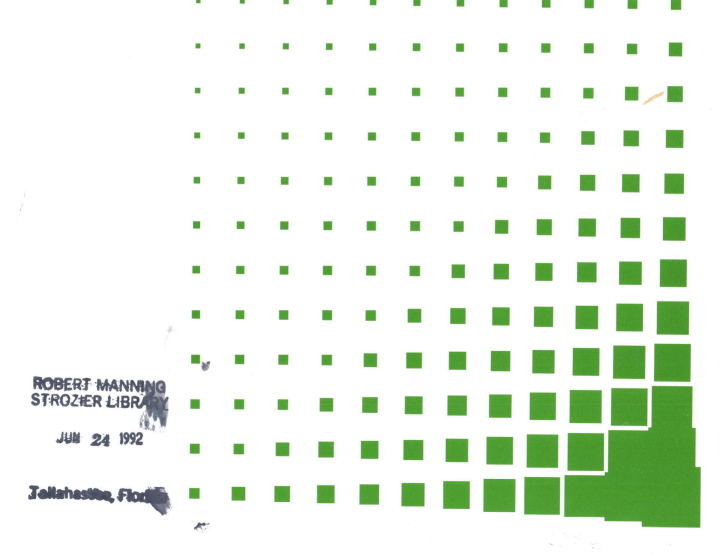

© Copyright, The Council of State Governments, 1989
ISBN 0-87292-091-7
C-141
$20.00

The Changing Arena: State Strategic Economic Development

CONTENTS

Figures

FOREWORD

The old rules of economic development may no longer apply. States are finding themselves playing in a new economic "arena" — one directly influenced by intergovernmental and international trends.

This report, by Lee Walker of the Division of Policy Analysis Services, focuses on the states' responses to this new global economic environment and the evolution of state strategic economic development. It includes a series of recommendations for states considering strategic policy action as a means of economic development in the 1990s.

The volume is the second in a series of three produced by Policy Analysis Services as a result of its study of state business tax and financial incentives.

The first, *State Business Incentives and Economic Growth: Are They Effective?* reviews the literature on the effectiveness of one economic development tool — business incentives — on state economic growth. The third report, *The States and Business Incentives: An Inventory of Tax and Financial Incentive Programs*, provides a state-by-state description of formal programs designed to create, expand and recruit business and industry.

Taken as a group, we believe these reports will provide an invaluable resource for state policy-makers considering various economic development options.

May 1989

Carl W. Stenberg
Executive Director
The Council of State Governments

ACKNOWLEDGMENTS

The author wishes to express appreciation to the states' economic development agency officials and the members of the CSG Corporate Associate Program for their time, effort and enthusiastic response to the CSG surveys used in this publication.

Many thanks to all of the authors and organizations listed in the bibliography for their work and contribution in the areas of economic development and public administration. Especially noteworthy are innovative ideas and writings by John M. Bryson, R. Scott Fosler, Roger J. Vaughan, Robert Pollard, Barbara Dyer, Jeffrey S. Luke, Curtis Ventriss, B.J. Reed, Christine M. Reed and Marianne K. Clarke.

And finally, a note of gratitude to CSG staff who contributed their thoughts and labor, including: Norm Sims, Debbie Gona, Keon Chi, Roger Wilson, Nancy Olson, Kevin Devlin, Lisa Brewer, Margaret Oberst, Jan Norris Clarke, Susan Stone, Pat Newsome, Daryl Theobald and Kathy Sutherland.

PREFACE

Business tax and other financial incentives used by states and communities to stimulate economic growth and create job opportunities are among the more controversial tools of economic development, particularly as a result of some of the headline-making incentive packages offered to auto plants and other large firms in recent years.

Proponents argue the use of incentives helps create a favorable business climate and that an attractive package of incentives is often the ammunition a state needs to compete in bidding wars against other states for industry. Opponents criticize the overuse of tax concessions and other fiscal inducements for their ineffectiveness and inefficiency, arguing that their costs far outweigh any long-term benefits.

State policy-makers face a fundamental challenge as they decide whether and how to tailor incentives to attract specific firms or to offer incentives as part of more comprehensive economic development strategies. To help them make more informed decisions, The Council of State Governments' (CSG) Policy Analysis Services undertook a study in 1988 related to state business tax and financial incentives.

Preliminary explorations of the issue, coupled with escalating media attention, suggested states *had* increased their use of incentives to attract business and industry. But what were the states offering in the way of formal tax and financial incentive programs? And were their numbers really increasing or were the headlines exaggerating the issue by grabbing on to some of the larger packages offered to firms?

More to the point, though, was there solid evidence that business incentives, as policy instruments, were effective in spurring economic growth in the states? Was the presence of business incentives even a factor in business investment and location decisions? If not, what forces were driving the expansion of their use? And what effect would the changing economic environment — one in which states are increasingly forced to compete in an international arena — have on the use of incentives and on economic development strategies in general?

The results of this undertaking, while revealing, do not offer any easy, uncomplicated answers for the states. However, they do offer a new look at some of the old assumptions surrounding the use of business incentives as a method for stimulating economic growth, as well as insight into some of the more recent economic development approaches and activities across the states.

Over the years, the number of states offering business incentives through formal programs *has* steadily increased. In fact, during the past five years a majority of the states offered a variety of tax and financial incentives for business and industry. Most states reduced their overall tax rates or tax rates for specific businesses. Some offered "customized" tax incentive packages for selected firms or projects. And to further assist businesses, most states initiated various financial incentive programs, apart from those already offered through existing federal programs.

And yet, a comprehensive review of past studies on the effects of incentives reveals *no* statistical evidence that business incentives actually create jobs. What those studies do suggest, overall, are some contradictory findings on the significance of incentives: they *are not* the primary or sole influence on business location decision-making and, relative to other factors, they *do not* have a primary effect on state employment growth; but they *do* become more effective when all other variables are equal among competing sites within a region or substate area, and they *are* important in that they often are used as a component in business climate indices.

Even though incentives cannot be linked to effecting tangible improvement in state economic growth, they apparently remain important "psychological" or political weapons. Some analysts argue that states have a common fear of being "outbid," and that states, in what amounts to an "arms race" mode of behavior, are forced into matching and beating each other's offerings, using incentives as defensive measures against their competitors.

But the field of "competitors" has expanded in the new economic environment. No longer is the competition just a neighboring state or region; now, it is national and international in scope, and states are responding to that new reality. For example, a 1988 CSG survey of state economic development agencies revealed that the packaging of business tax and financial incentives remains an important issue and that states believe those inducements can have a significant effect on new business investment and job creation. Increasingly, however, states are using incentives within the context of a comprehensive economic development strategy that accounts for their strengths and weaknesses with regard to regional, national and international competition.

Indeed, the new global economic environment that has emerged over the last decade is limiting the states' capacity to pursue economic development in older, more conventional ways. The changes that created the economic restructuring of the 1980s are likely to accelerate, and states will need to reconsider their approaches to economic development and adjust their policies to survive and prosper in the world economy. That is the reality for state policy-makers as the decade of the 90s draws nearer.

This is just one volume in a series of three companion pieces produced by Policy Analysis Services as a result of its 1988 study: *State Business Incentives and Economic Growth: Are They Effective? A Review of the Literature; The Changing Arena: State Strategic Economic Development;* and *The States and Business Incentives: An Inventory of Tax and Financial Incentive Programs.*

Each provides insight into one piece of this economic development puzzle. Together, they offer a set of perspectives on where the future is likely to lead the states in their economic development efforts and in their interactions with each other and the rest of the world.

Deborah A. Gona
Director, CSG Policy Analysis Services

> *Americans today face the sobering question of whether, in the perspective of history, their economic leadership will turn out to have been a relatively brief period of less than one century, during which world economic leadership passed through North America from one group of economic giants in Europe to another in Asia. Each American state is inescapably a part of that national story and the worldwide forces that have molded it.*
>
> R. Scott Fosler
> *The New Economic Role of the American States:*
> *Strategies in a Competitive World Economy (1988)*

The Changing Arena:
State Strategic Economic Development

States have been active in economic development since the 1930s. But only within the past decade has development emerged as a critical function of state government. New federal-state relationships in domestic policy and the emergence of a global economy have had a profound impact on state economic development policies in the 1980s. This report examines current state economic development and specific policy considerations that are changing the way states define their economic development roles.

The examination of current state economic development begins with a discussion of traditional state economic development policy-making and its evolution during the 1980s. The policy question that emerges is how can states, in a cost effective way, generate businesses and jobs that serve the public good?

In response to that question, the report addresses the nature of economic development policy-making and a variety of policy considerations that are prevalent in today's state economic environment, including the new domestic and world view. The states' responses to the changing economic arena, including the movement toward more collaborative and integrated approaches to economic development; their use and assessment of business incentives; and state governments' and businesses' need for reliable information and for collaboration between the public and private sectors are discussed within the new context of the states' global economic development environment. The global environment and the private sector and state government response to it, are described as they have evolved in the 1980s — particularly in the form of state strategic economic development — the combination of strategic planning and

Overview

The global economic environment that has emerged in the 1980s limits state governments' capacity to pursue economic development in older, conventional ways. Increasing national and international competition has hampered industrial recruitment strategies and altered traditional assumptions of what is required to stimulate private investment. As participants in a new intergovernmental, intersectoral arena, states are now compelled to adopt collaborative strategies, since they can no longer deal with economic problems independently.

The tremendous changes that created economic restructuring during the 1980s, will accelerate in the 1990s. State governments will need to be much more flexible, responsive and sophisticated in order to even minimally serve their constituents, and state economic development policy will have to be adjusted to address these realities. States will need to develop economic development policies that are comprehensive, collaborative, future-oriented and pro-active if they are to survive and prosper in the world economy.

action as an approach to public policy-making. A significant outcome of this discussion is the evolution of the original policy question into the policy issue of state governments' development of economic capacity and their creation of wealth to generate businesses and jobs that serve the public good in a cost-effective way.

The report concludes with a state economic development policy forecast for the 1990s, including an evaluation of state strategic economic development and recommendations for states considering the use of strategic policy action as a means of economic development. The policy analysis, recommendations and specific graphics included in the report are intended for use as helpful and informative points of reference for state economic development policy-makers.

Incentives should be tailored for very specific strategic purposes. They should not have as their sole purpose zero-sum moves from State A to State B simply because B is willing to forego all business taxes for a few jobs.

Arthur Young & Co.

Current State Economic Development Policy

To understand what state economic development is and what it might become in the 1990s, we must first consider traditional state economic development policy-making and its evolution through the 1980s. A variety of factors — including the policy-makers themselves, the obstacles to policy-making and the state government and private sector information gap — have played key roles in shaping current state economic development policy.

The Tradition of State Economic Development

State economic development activities have evolved largely as pragmatic responses to unemployment. In order to relieve the economic, social and political stress it creates, state development decisionmakers have acted generally on the assumption that jobs are good, no matter what their source, and more jobs are even better.

The recruitment or relocation of companies, particularly large ones, became a popular economic development activity during the 1950s and expanded substantially during the 1960s and 1970s. Conventional recruitment programs offered tax concessions, business loans and other financial incentives, job training grants, infrastructure improvements, site selection assistance and various forms of business-specific assistance. But manufacturing employment in the United States declined from 1972 until the recession of 1983, and studies began to indicate that most job growth did not result from the relocation of industry, but rather from the expansion of existing firms or the creation of new firms.

Critics charged that state economic development based on business recruitment essentially was a zero-sum game from the national perspective. The process of inducing firms to relocate their facilities in areas where governments wanted them to really did not create jobs, it simply moved them. However, economic development decisionmakers, with the mission of alleviating unemployment in their respective states, were under pressure to generate jobs, regardless of other states' losses.

At the same time, however, states also began to realize that they were losing jobs faster than they could be replaced through traditional recruitment methods alone. In reaction, states focused new efforts on the retention of existing businesses through tactics such as unemployment insurance, workers compensation, plant closing legislation and job training, as well as direct subsidies to businesses.

Unfortunately, state-sponsored job retention efforts generated their own set of problems. Some of the retention measures proved unsuccessful, costly or even counterproductive. Moreover, as some states have since learned, certain businesses and jobs are not necessarily good targets for retention, particularly if they impede the state's progress in acquiring new, more competitive industries and viable jobs.

Nevertheless, competition between states for businesses and jobs increased noticeably during the 1980s, and states increased their use of business incentives. That competition, in turn, raised the minimum level of incentives states felt they had to offer. Paradoxically, the competitive difference between the states actually narrowed since many of the same incentives were being offered. As a result, the relative impact of these government subsidies per se declined, although they continue to be important — but not necessarily accurate — indicators by which private sector observers evaluate the states' business climates.

The predicament for the states, though, has been both an economic and a political one. State executives and legislators have felt pressured to use incentives as defensive measures to prevent and compensate for the loss of jobs to other states and to address unemployment and economic development issues by generating employment opportunities. Even states that have resisted participating in this form of economic competition have felt, and will continue to feel, the impact, either by increased subsidy costs or unfavorable evaluations in business climate rankings.

That is the current paradox. If states stop offering incentives, no matter how cost-ineffective, they will be adversely affected in terms of their business and economic image and, subsequently, their ability to generate jobs. The policy question arising from this predicament then is how can states, in a cost-effective way, generate businesses and jobs that serve the public good? In order to address that question, it is first necessary to review the process of state economic development policy-making as it currently exists.

The Nature of Economic Development Policy-Making

State economic development policy is created through complex, interdependent relationships between governors, legislative leadership, state economic development agencies, development corporations, business groups and a host of interested parties, including other public agencies and private groups. Because no single agency exercises comprehensive control over economic development efforts, gaps, overlaps and lack of coordination are inherent in the process. (See State Actions Critical to the Process of Economic Development, p. 4.)

The policy-makers

Governors increasingly have played a central and critical role in establishing economic development policy and in setting the scope of activity. As political leaders and managers of state government, they are in a unique position to bring together a variety of actors to structure economic development policy and build coalitions to support economic initiatives. They are able to play a key role, for example, in involving private industry, universities and local governments in the states' economic development efforts, along with a variety of public agencies whose activities affect the state economy.

Governors and their administrations often play highly visible roles in promoting their states for business and economic development. Advertising campaigns, trade missions and the distribution of brochures depicting the states' demographics, labor force, infrastructure, education systems and business support programs are used by most governors. However, since economic concerns are almost always a high public priority, gubernatorial involvement in economic development has become as much a pragmatic, political response as a policy preference.

One-stop shopping in state development offices, who bring together all the folks we will need to deal with, is important.

The E.I. duPont Co.

State legislators, particularly legislative leadership, are similarly involved in state economic development policy-making. Legislative influence comes through the enactment of laws, the adoption of budgets, the approval of expenditures and the establishment of taxes directed toward specific businesses and industries or public agencies that regulate the state business and economic environment. Generally, legislators play a more reactive role, one dependent on the extent of their opportunity to collaborate with their governors in determining economic development policy.

State economic development agency officials typically are the states' specialists in business recruitment and economic development policy implementation. As technocrats, they are well-versed in their state's business and economic laws, regulations, taxes and resources. As officials in an executive branch agency, one usually closely associated with the governor,

State Actions Critical to the Process of Economic Development

Human Resources. The skill, motivation, cost and adaptability of the work force is affected by state programs in primary, secondary, vocational, community college and higher education; job training; employment service, unemployment insurance and workers' compensation; income maintenance and welfare; health and human services; and labor relations (including right to work, collective bargaining, and strike laws).

Physical Infrastructure. The network of facilities required to conduct business and commerce includes infrastructure that is financed, constructed, operated, maintained or regulated entirely or in part by the state. These include transportation facilities (roads, bridges, mass transit, ports, airports, railroads, etc.), water supply and sanitation, solid waste disposal, communications, energy and housing.

Natural Resources. States regulate or directly manage key natural resources that either may be the direct basis of business enterprises or indirectly affect business activity by constituting part of the "quality of life" or tourist attractiveness of the state. These include land (both space and soil), water (for supply, industrial use, transportation, seafood, recreation, etc.), air, agriculture, minerals, forests and wildlife.

Knowledge and Technology. States are major producers and disseminators of knowledge and information and supporters of research and the development of new technology. They finance public universities and research institutions, and they promote links between businesses and knowledge-based institutions in order to encourage the commercialization of research products.

Enterprise Development. State actions directly affect the organization, financing, location and operation of business enterprises. These include programs designed to encourage the start-up, expansion and attraction of business through financial and technical assistance, business incubators, research parks, enterprise zones and export promotion. Included here are the conventionally defined "economic development" activities, which have become far more sophisticated and selectively targeted in some states.

States also directly affect enterprises through regulatory activities that cut across every phase of business activity. Especially important is the regulation of private financial institutions (banks, savings and loans, insurance companies, etc.) States also provide capital directly through seed, risk and expansion financing mechanisms.

Quality of Life. The general quality of life affects the economy in two principal ways: it is a direct source of business enterprise (e.g., tourism, travel, recreation, leisure); and it is an important factor in attracting and retaining businesses and workers. States affect the quality of life through most of their actions in other areas but, in general, by providing good public services, directly providing or encouraging the private provision of other desirable public amenities (e.g., hospitals, educational institutions, museums, cultural activities, etc.) and assuring an attractive and healthy physical environment.

Fiscal Management. State taxes, fees and user charges affect both the cost of doing business and the personal expenses of employees. The structure of taxes (i.e., who is taxed, how much, on what basis) can affect business decisions regarding start-up, investment, innovation, expansion, contraction and relocation. The revenue base also determines the ability of state government to finance activities in the first six categories.

Source: R. Scott Fosler, The New Economic Role of the American States: Strategies in a Competitive World Economy (1988).

they most often are the intermediaries between government and business. In those capacities, they are important information providers to the business community, legislators, local governments and other public agencies that may be affected by economic development activities.

Increasingly, the business community — in the form of advisory councils, private organizations and community groups — has become an active participant and collaborator in state economic development policy-making. Many government and business leaders now are openly acknowledging the interrelatedness of the sectors and the importance of these new public-private partnerships to effective state economic development policy-making.

Institutional obstacles to policy-making

However, bureaucratic structures, short-sighted legislation, miscommunication between government and business and fierce competition for profits continue to be major obstacles to effective economic development policy-making in the states. At the agency level of state government, the standard operating procedures, federal and state program regulations and interest group and clientele demands make it difficult for state economic development officials to commit the necessary time and resources to overall policy development. As a result, these officials more often are compelled to react to daily demands instead of progressing toward more far-ranging visions of economic development.

In addition, state legislation that is not carefully crafted may be self-defeating. Laws that limit competition, attract industry, encourage new business formation or regulate existing businesses have a significant impact on each state's business community and economic environment. Recognizing that, many government and business leaders now are making an effort to collaborate more often in formulating legislation that affects economic development, whether it pertains to business, education or the environment. Such public-private partnerships, though increasingly effective, are constantly strained by the conflicting pressures of profit motive and public responsibility. Yet, there has been a growing realization that the inherent communication gap between government and business must be eliminated if states are going to be suc-

cessful in their economic development efforts in the 1990s.

The information gap

Successful communication between government and business is largely dependent on the quality of information that each provides for the other. The availability of quality information on state business environments is especially important in securing effective public-private partnerships for economic development policy-making. However, many business leaders are unable to obtain current, accurate information regarding the states' business environments, or climates, as they are most frequently characterized. And state government leaders are equally concerned about their inability to obtain useful information on business needs and priorities — information that can be used to help them make informed location and expansion decisions.

> *Underlying the specific elements of a favorable business climate is the need for stability and consistency. A current favorable climate is meaningless if there is a history of constant change.*
>
> The Procter & Gamble Co.

Existing information, particularly as presented in business climate reports, has not proven reliable to business or government decisionmakers. During the 1970s, "business climate" became the key phrase for an aggregate evaluation of various quantifiable and intangible factors that could affect, either favorably or unfavorably, the conduct of business in a particular state. The business climate concept provided a way to rank each state's competitive position or comparative advantage in terms of its attractiveness as an environment in which to do business. The private firms that conduct the most widely recognized business climate studies — such as Grant Thornton, the Corporation for Enterprise Development (CfED), *Inc. Magazine* and SRI/AmeriTrust — have published their findings in the form of national reports in which each state is ranked according to the quality of its business climate.

Increasingly, these reports and rankings have been criticized by government and business leaders for oversimplifying or even ignoring crucial factors. The Grant Thornton report, for example, focuses almost exclusively on the costs of doing business, while ignoring education and the state's regulatory environment, two of the important public policy factors that may enhance a state's attractiveness as a business location. The 1987 CfED study examined job and income growth, entrepreneurial activity, labor force skills, technological innovation, capital availability and quality of life amenities, as well as conventional cost factors in determining the quality of a state's business climate. For its annual business climate ranking, *Inc. Magazine*, uses job growth and new business formation as measures to determine entrepreneurial activity, which is used almost exclusively to determine each state's business climate.

> *Reliability and predictability in government policies and actions are the most critical site-selection needs. HP [Hewlett Packard] has facilities in high-tax states, but it's fairly comfortable in those states because expectations generally equal reality.*
>
> Hewlett Packard

State government leaders complain that these reports use outdated and incomplete information, resulting in inaccurate and misleading rankings. Even reports that include large amounts of data, they maintain, are largely subjective, while others, presenting aggregate data, simply fail to take into account state-to-state variances in economic conditions and business criteria. These studies, their critics say, focus on the development of static profiles or measures of the business climate, even though trend data would more likely provide accurate and insightful indicators of a state's business and economic conditions. Furthermore, economists and business leaders are beginning to agree that the emphasis no longer should be on each state's business climate, but rather on each state's economic capacity.

Similarly, at the state level, limited information and superficial analyses often weaken policy-making efforts. For example, compared to the way firms make location decisions, state governments undertake relatively limited analyses to determine which and how many incentives should be offered as inducements to attract particular businesses. And they give less attention to cost and benefit considerations than the corporate side. Generally, the private sector operates with a clearer vision than government when it comes to negotiations for location or expansion (see The Site Selection Process, p. 7). More often than not, business negotiators know what government must provide in order for their firms to operate profitably. State economic development officials, on the other hand, are compelled to calculate each project's impact in terms of the public good, an equation with considerably more variables than gross or net profits.

However, many state policy-makers have not considered the full scope of the costs of location or expansion inducements. Consequently, in the case of business incentives, for example, they focus attention almost exclusively on the benefits, with the most common assumption being that the increased revenues from each project will allow it to pay its own way. However, even if a new project can generate revenue sufficient to cover the costs of the incentives used to attract it, the project still may be costly to the state's taxpayers. Dramatic population increases, overloaded transportation networks and strained public services may necessitate tax increases and lead to a diminished quality of life, simply because these considerations were not factored into the development scenario.

Policy Considerations in a Changing Arena

Many policy considerations must be taken into account in order to determine the best approach to state economic development. Business and economic climates, educational systems, infrastructure networks and environmental concerns encompass a multiplicity of micro-economic and social factors. Although the best approach will vary from state to state, some fundamental factors are generic to economic development policy-making and policy-execution in the states.

The economic environment, the exchange of information between government and business and the types of development activities that states can employ continually come to bear on state economic development decision-making. For decades, each of these has influenced enormously the shape and content of state economic development policy. However, during the 1980s, the economic environment has been the most volatile factor of the three.

The Economic Environment

Among the many facets of the economic environment, two particular phenomena are drastically altering state economic development policy. The first is the change in the relationship between the federal and state governments with regard to domestic policy. The second, and more dynamic development, is the emergence of a truly global economy, one in which each state has become an active participant (see The Global Economy, p. 9). The combination of these two events is forcing each state to rethink its approach to economic development and recast its role in the marketplace and the

The Site Selection Process

The locational decision is part of a larger corporate planning process. The organization of the company determines the structure of the site selection team. Corporations with a centralized staff will generally form a team including representatives from key areas such as transportation, distribution, personnel, engineering, real estate and planning. Companies with strong divisions and weaker corporate staffs may carry out the locational studies at the divisional level. In this case, the corporate office would normally retain supervising authority.

The site selection team will develop a list of characteristics that are important for the location of the new facility. A "must-and-want" list will include both quantifiable and nonquantifiable locational factors. The locational factors may be weighed to indicate which locational features are most important. The next step is to gather information about potential sites to compare the features of each site against the "must-and-want" list. Sites will be eliminated in rounds as more detailed and difficult to obtain information is gathered after each elimination round.

The locational decision is normally made sequentially. The first stage is the choice of a state or region. Over half of all locational studies make their first cut at a multistate level. Once the state or region has been selected, a more micro-focus will culminate in the selection of a specific community and site. The important locational factors differ between the first stage when firms are seeking a general region in which to locate and the second, more geographically focused stage. In selecting a broad region, the site selection team will focus on labor, state

tax variables, climate, proximity to markets and other features that may have significant interregional variation, but are similar almost everywhere within the region. Locational factors that vary at the micro-geographic level of detail such as land costs, access to major roads and schools are generally available somewhere within all major regions. Micro-factors become more important when locational studies do not distinguish between stages of the locational process. As the locational choices are narrowed, discussions with local public officials regarding potential problems and incentives begin. A feasibility analysis must normally show that the proposed plan will earn a high enough rate of return to justify the construction costs.

Recent studies reveal that industrial location choices are governed to a lesser extent than in the past by access to markets, labor, transportation and raw materials. These traditional location factors still exert an important influence, however, the list of important locational determinants has been expanded to include state and local taxes, education, business climate, labor, skills, and state and local physical infrastructure. At the high technology end of the industrial spectrum, these nontraditional location factors tend to dominate the location choices. As the nation's economy continues to shift to advanced technologies to remain competitive, the overall importance of the nontraditional location factors will increase as the traditional factors will continue to be quantitatively more significant in terms of their overall influence.

Source: John P. Blair and Robert Premus, "Major Factors in Industrial Location: A Review," Economic Development Quarterly (February 1987).

world economy. As a result, most states are developing a heightened awareness of the micro- and macro-economic factors that influence their economies, are becoming increasingly pro-active in international business and finance and are becoming much more amenable to public-private partnerships in economic development projects.

The alteration of state economic development policy — a new domestic and world view

The shift in the domestic policy relationship between the federal and state governments that began in the 1970s during the Carter administration and accelerated during the Reagan administration represented a major federal withdrawal on the domestic front — one that created a fundamental alteration in conventional state economic development roles. The result has been a shift in domestic responsibilities and a reactivation of state powers to address issues that previously were within the purview of the federal government. States now are in a position to exert more direct influence on their own economies. Conversely, they are in a position to feel directly — in terms of revenue and employment rates — the effects of the successes or failures that result from their economic intervention.

That states are responding to economic demands with interventionist measures, however, does not automatically eliminate the influence the federal government has on each state's economy. Through massive expenditures, the trade and exchange rates and fiscal and monetary policies, the federal government continues to exert substantial influence on the national economy and, therefore, on each state's economy. Likewise, each state, to varying degrees, feels the effects of industry-specific federal regulations, trade deficits and the national debt. It can be argued that such factors offer a compelling case for states to assume pro-active roles in order to stabilize and improve their economies.

An even more compelling reason, however, is the emergence of a global economy and the ramifications it presents to every state in economic, political and social terms. During the 1970s, technological advances and world market transforma-tions contributed to the increasing vulnerability of the U.S. economy to foreign competition. Throughout the 1980s, as states evolved into their new roles as economic interventionists, their active involvement in world markets also increased. The dynamics of the last two decades have resulted in an economy, global in scope, in which states are individually and collectively more sensitive to world market conditions. State governments' approaches to economic development in the 1990s will be largely determined by these realities.

States' Response to the New Agenda

Throughout the 1980s, the states' aggressiveness in addressing the new economic agenda has been demonstrated by the growing budgets of state economic development agencies — budgets that have increased dramatically during the decade, according to surveys conducted by the National Association of State Development Agencies. During the same time period, federal economic and community development expenditures decreased substantially, as reported by the U.S. Office of Management and Budget.

To determine exactly what state governments and businesses currently are thinking and doing with regard to development activity, The Council of State Governments (CSG), in the summer and fall of 1988, surveyed the states' economic development agencies, as well as members of the private sector through the CSG Corporate Associates Program. The responses provide useful insight not only toward understanding how states view the use of business tax and financial incentives, but also how their economic development efforts reflect their changing role in the world economy.

The responses from the economic development agencies indicate that state governments are moving toward more collaborative and integrated approaches to economic development. Comprehensive economic development plans currently are in use by 31 states (see Figures 1 and 2), and 17 of those are employing strategic approaches to economic development (see Figure 3). It is important to note, however, that of the states not currently using comprehensive plans or strategic approaches, a significant number are southeastern states, many of which have relied heavily on industrial recruitment as the driving force in their economic development activities. It also is significant that four of those southeastern states, along with seven states in other regions, currently are preparing comprehensive plans, another indication of the growing trend toward approaching state economic development in terms of a more comprehensive strategy.

The continuing use of incentives

However, an overwhelming majority of states still consider the packaging of business tax and other financial incentives to be an important issue in their economic development efforts (see Figures 4 and 5). Most of the survey respondents indicated there has been more pressure on their states to offer a better set of incentives than others in order to remain competitive (see Figures 6 and 7). And, furthermore, most feel that business tax and other financial incentives can have a significant effect on new business investment and job creation in their state (see Figures 8 and 9).

Although business tax and financial incentives are not the primary focus of most states' economic development efforts, they remain important parts of state economic development strategy, particularly in

---

The Global Economy

New Knowledge. The primary force driving the world economy today is the flood of new knowledge pouring out of the world's research laboratories.

Internationalization. A major trend is the shift from self-sufficient national economies to an integrated system of world-wide production.

Capital Movement. Capital movements rather than trade have become the more predominant force in the world economy. The economy of goods and services and the economy of money, credit and capital are no longer bound tightly to each other. They are moving in different directions and at different speeds.

Stock Around the Clock. Stock trading through computers is creating a new generation of financial interchange. With people using their personal computers to get real-time data, the pulse of stock trading has been quickened and the volume increased.

America Is Not Deindustrialized. What is changing is that the increased use of sophisticated technology is reducing employment in manufacturing industries.

Japanese Challenge. Japan is on the way to becoming the world's premier investor and creditor. Japan's economic and financial clout could eventually rival the power held by Britain in the nineteenth century and the United States after World War II.

Privatization. One of the distinguishing features of world economic activity is the dismantling of centrally controlled production and the fostering of private enterprise. As the world moves toward the twenty-first century, it becomes increasingly clear that a centrally planned economy with its rigidities and lack of competition, cannot succeed in a world economy characterized by interdependence, rapid technological innovation, and the need for swift decisions.

Soviet Joint Ventures. The Soviet Union is exploring the possibility of joint ventures with Western European and Japanese firms. Russia may thus secure Western technology without drawing on its dwindling supply of hard currency.

Corporate Restructuring. With change pulsing through the manufacturing system so fast, every major corporation in America is being forced to redefine what business it is in. Technology is changing so quickly that companies do not necessarily want to tie themselves down to making a particular product. Joint ventures and other alliances offer far more flexibility in a fast-changing environment.

Source: William Van Dusen Wishard, "The 21st Century Economy," The Futurist (May/June 1987).

light of foreign investment in the states. For example, by 1986, at least 40 states had foreign offices. In Kentucky, 49 foreign manufacturers have located new facilities in the state since 1983. North Carolina has designated Foreign Trade Zones strategically located across the state to offer cost advantages to those businesses engaged in overseas trade. And in 1986, the New Jersey Department of Commerce and Economic Development reported that more than 1,000 foreign firms from 48 different countries provided jobs for about 160,000 of that state's residents.

States' economies are becoming more integrated into the global economy, which means that state economic development policies are being guided more by world market and international economic forces than by domestic political initiatives. States continue to use tax and business incentives, for example, but usually within the context of a comprehensive economic development strategy that accounts for the states' strengths and weaknesses with regard to regional, national and global competition.

Assessing the costs and benefits

As a result, most of the survey respondents indicated that their states have initiated the development of guidelines or a set of specific criteria to be used in evaluating businesses wishing to receive tax or other financial incentives (see Figure 10).

Such guidelines or criteria are important points of reference for state economic development officials and provide specific direction for state comprehensive or strategic economic development policy-making. For example, the calculation of costs and benefits associated with packaging tax and financial incentives for businesses has been a difficult, if not impossible, task for states trying to determine how to invest in economic development. Although such cost-benefit equations are critical components of the guidelines or criteria that are included in comprehensive plans and strategic ap-

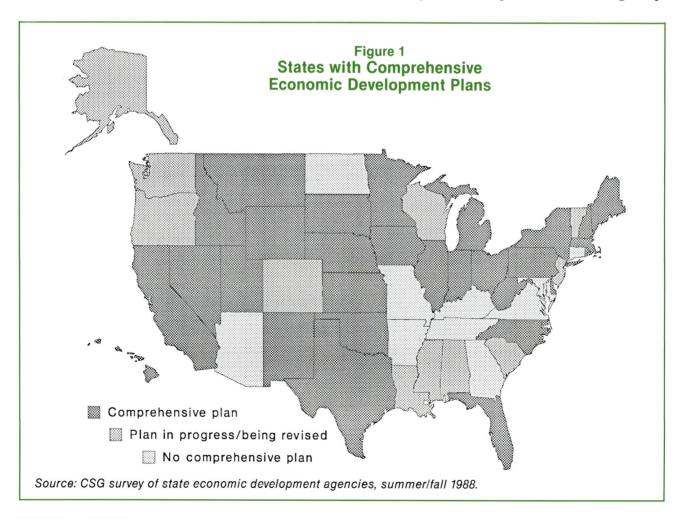

Figure 1
States with Comprehensive Economic Development Plans

Comprehensive plan

Plan in progress/being revised

No comprehensive plan

Source: CSG survey of state economic development agencies, summer/fall 1988.

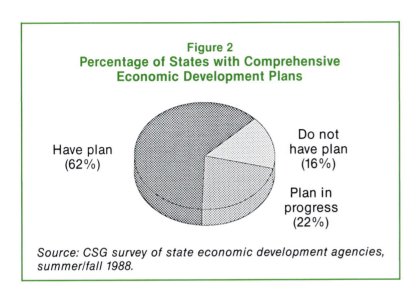

Figure 2
Percentage of States with Comprehensive Economic Development Plans

Have plan (62%)

Do not have plan (16%)

Plan in progress (22%)

Source: CSG survey of state economic development agencies, summer/fall 1988.

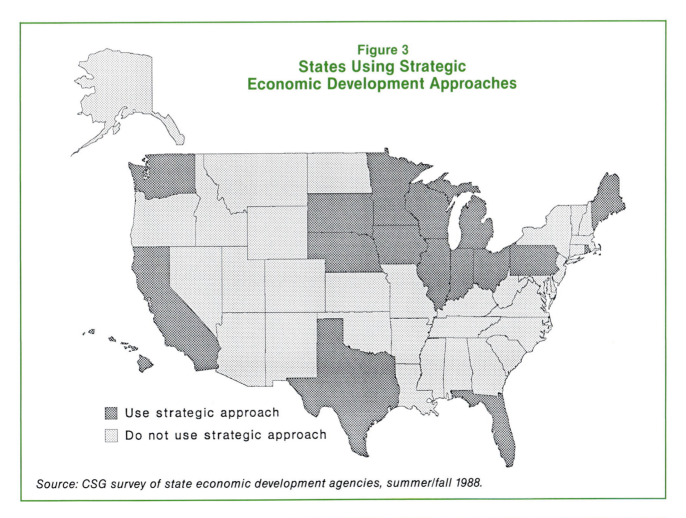

Figure 3
States Using Strategic Economic Development Approaches

Use strategic approach

Do not use strategic approach

Source: CSG survey of state economic development agencies, summer/fall 1988.

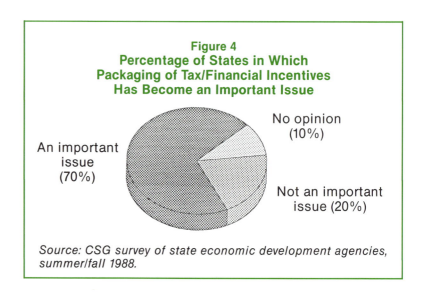

Figure 4
Percentage of States in Which
Packaging of Tax/Financial Incentives
Has Become an Important Issue

No opinion (10%)

An important issue (70%)

Not an important issue (20%)

Source: CSG survey of state economic development agencies, summer/fall 1988.

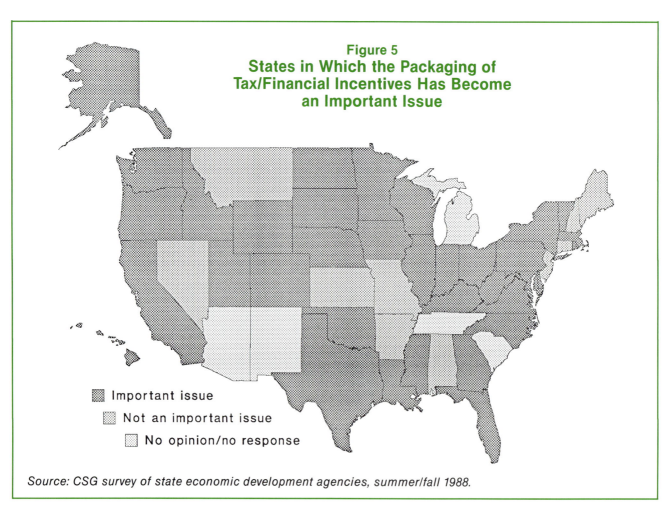

Figure 5
States in Which the Packaging of
Tax/Financial Incentives Has Become
an Important Issue

Important issue

Not an important issue

No opinion/no response

Source: CSG survey of state economic development agencies, summer/fall 1988.

proaches, many states continue to struggle with the complexities of cost-benefit analyses, both for specific projects and overall strategies.

Unfortunately, the survey responses also suggest that many states have a limited capacity to perform in-depth economic development cost-benefit analyses (see Appendix A). Most that do perform these analyses rely on state economic development agency calculations of jobs created and state dollars invested or on legislative review of appropriations and tax impact. Both, however, tend to be project-specific. Comprehensive and strategic approaches, on the other hand, call for a much broader consideration of costs and benefits, taking into account both the immediate and long-range impact of economic development programs and policies on education, infrastructure, social services, state fiscal capacity and the environment. Some states, notably those using strategic techniques, already have begun to evaluate projects, programs and

policies from a broader perspective, one that acknowledges the importance of other areas related to economic development. However, given the many public costs and benefits that result from economic development activity, the majority of state economic development agency respondents indicated they need better information and more reliable methodologies if they are going to be able to accurately assess the impact of their states' projects, programs and policies (see Appendix B).

The need for reliable information

Better information, both coming into and going out of the agencies, along with better information delivery systems — including communication infrastructures, educational systems and current data bases — are critical needs of almost every state. Traditionally, these information needs have been fundamental to the states' development activities. How they have been met has largely determined

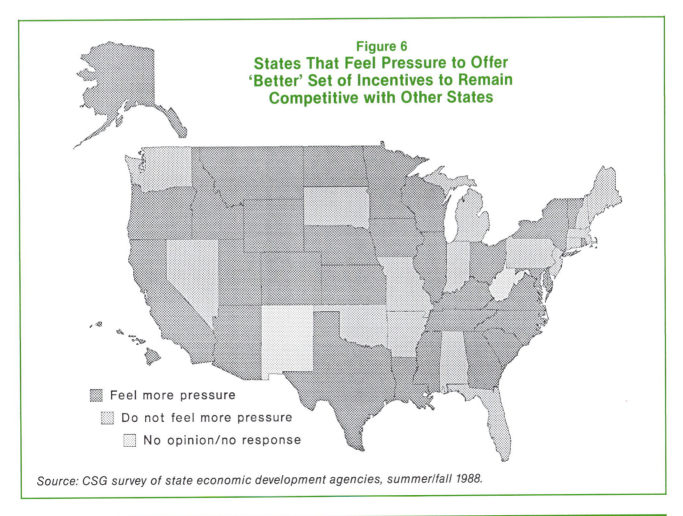

Figure 6
**States That Feel Pressure to Offer
'Better' Set of Incentives to Remain
Competitive with Other States**

▨ Feel more pressure
▦ Do not feel more pressure
▦ No opinion/no response

Source: CSG survey of state economic development agencies, summer/fall 1988.

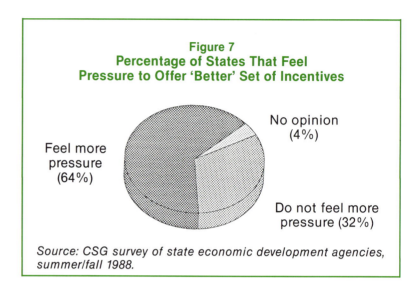

**Figure 7
Percentage of States That Feel
Pressure to Offer 'Better' Set of Incentives**

No opinion
(4%)

Feel more
pressure
(64%)

Do not feel more
pressure (32%)

*Source: CSG survey of state economic development agencies,
summer/fall 1988.*

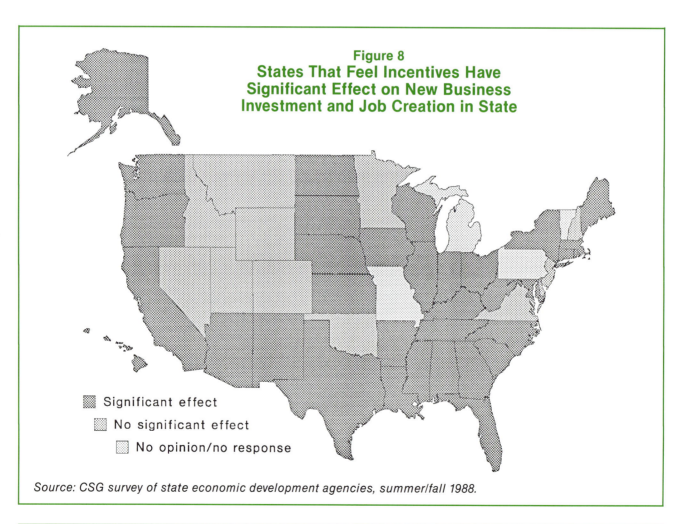

**Figure 8
States That Feel Incentives Have
Significant Effect on New Business
Investment and Job Creation in State**

Significant effect

No significant effect

No opinion/no response

Source: CSG survey of state economic development agencies, summer/fall 1988.

how successful the states have been in competing with one another. Today, how those needs are met determines whether or not a state maintains the economic capacity to survive in a global market. State governments need reliable and comparable data regarding markets, industries, business costs, economic forecasts, emerging technology, labor demographics and innovative economic development programs. Just as importantly, they require systems to develop and communicate that information on an on-going basis.

The lack of good information, in fact, was identified by most of the state economic development agency respondents as the biggest obstacle to their efforts to encourage businesses to locate or expand in their respective states (see Appendix C). These agencies identified misinformation about their state's business climate, general misconceptions about the state, inadequate information-sharing among government agencies and between the public and private sectors and the absence of dependable delivery systems offering comparable data as major obstacles to their economic development efforts.

The respondents from the CSG Corporate Associates Program also pointed to unreliable data and unpredictable information sources as hindrances to their successful interaction with state government officials. It is important to these firms that reliable information be available regarding the states' regulatory environment, tax codes and policies, labor force, business assistance programs and key infrastructure systems, such as transportation, communication and education. Despite their generally productive relations with state economic development agencies, the respondents maintained that state business climates would be more favorable if more current and complete data were provided to them and if state tax and regulatory policies were more consistent and equitable. A favorable business climate, according to these private sector respondents, is both stable and predictable — conditions that consistent state economic development policies help create.

Collaboration between public and private sectors

As they face the turbulent challenges that world markets and the global economy present, businesses and industry are seeking stability and clarity in state government policies and procedures. It is a situation in which the public and private sectors are becoming increasingly dependent upon one another, and in which the potential for substantive collaboration exists. The extent to which state governments are able to provide more efficient and sophisticated economic environments, and the degree to which business and industry accept roles as responsible corporate citizens determines the frequency and success of this collaboration.

> *You have to know what the rules are and have a reasonable level of confidence that the rules will not be changed dramatically.*
> Blue Cross/Blue Shield Assn.

The likelihood that such public-private sector collaboration will be successful has been increased with the development of a common perspective or view of the economic world. State governments and businesses, alike, are beginning to recognize that the global economy demands flexibility, responsiveness and quality communication, not just to prosper, but to survive. As a result, during the 1980s, both sectors have turned increasingly toward strategic planning, management and action as a means of meeting these new demands.

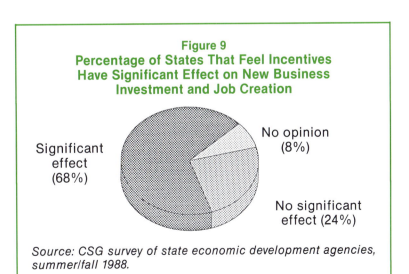

Figure 9
Percentage of States That Feel Incentives Have Significant Effect on New Business Investment and Job Creation

Significant effect (68%)

No opinion (8%)

No significant effect (24%)

Source: CSG survey of state economic development agencies, summer/fall 1988.

Strategic Action in the Global Economy

The blurring of the boundaries between public, private and nonprofit sectors has created an interrelated, global economy in which institutions are highly influenced by one another. Each participant is compelled to execute innovative strategies for success and survival. In response, many state governments have adopted strategic planning and management as one means of becoming strategically active as interventionists in today's and tomorrow's global economy.

As concepts, strategic planning and management have been parts of planning education since the 1960s (see Figure 11). In fact, many state administrators may view what they perceive to be as another management technique with skepticism. They have seen cost-benefit analysis, planning-programming-budgeting systems, zero-based budgeting, management by objectives and other management novelties fall by the wayside after en-thusiastic proclamations by their designers, various authors and neo-believers. Their skepticism would not be entirely unwarranted if strategic approaches are limited to public agency management as opposed to public policy-making. In either case, caution must be exercised to tailor strategic approaches to serve specific purposes and situations. A variety of approaches to strategic planning and strategic management — each based on a diagnosis-vision-action process — exist. However, these approaches are not all equally applicable to state government (see Appendix E).

Strategic planning requires broad-scale information gathering, an analysis of alternatives and a focus on the future impact of decisions. It requires an organization to deal with critical emerging and future issues by way of a strategy, long-term objectives and integrated programs for accomplishing these objectives. For public organizations, strategic planning addresses the political, economic and legal factors

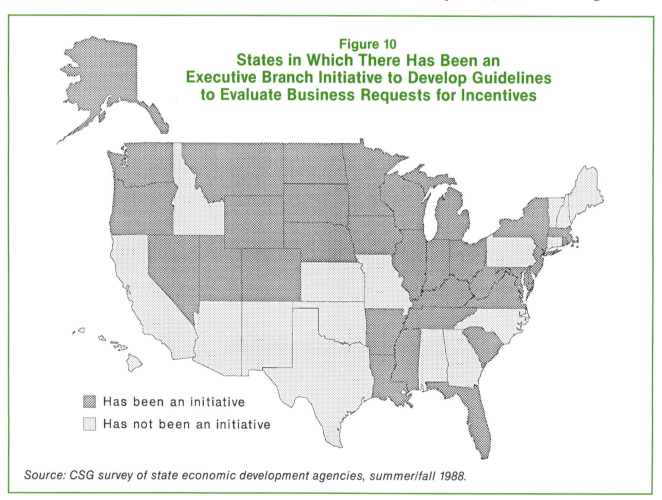

Figure 10
States in Which There Has Been an Executive Branch Initiative to Develop Guidelines to Evaluate Business Requests for Incentives

Has been an initiative

Has not been an initiative

Source: CSG survey of state economic development agencies, summer/fall 1988.

that shape public policy, and can accommodate divergent interests and values by facilitating communication and participation (see Appendix F).

Strategic management is a frame of mind and mode of behavior — one that accepts change as a permanent condition and calls for strategic adjustment as the market demands it. Ultimately, it is synonymous with creating change, rather than simply reacting to it. Strategic management incorporates the fundamental processes of strategic planning, foresight, goal setting, systematic linking between strategic plans and operations and evaluation. The strategic management approach to economic development requires states to analyze their economic, political, human and natural resources in order to identify and continuously evaluate long-range policy actions and goals that will best create economic wealth commensurate with a state's strategic strengths. Ideally, action, results and evaluation will occur continuously, not just at the end of the process.

Most strategic planning and management have occurred in the private sector and have been concerned primarily with profits and markets. Constant change and complex markets demand the ability to think strategically and the flexibility to react quickly to market conditions. As public sector entities, states operate within specific governmental and political systems and are less market-oriented. For state governments, any strategic action must necessarily be concerned with constitutional arrangements, legislative and judicial mandates, government rules and regulations, jurisdictional boundaries, resource constraints, political climate factors and client and constituent interests. Moreover, strategic action by government agencies necessarily includes external demands, constraints and mandates with agency-specific missions and operational procedures. Effective strategic planning and management at the state level, therefore, must recognize and accommodate the related political dimensions of those factors.

This dynamic interdependence, particularly as it exists within the context of rapidly advancing communication networks, is a critical consideration for economic development policy-makers. Strategic planning and management, with their emphasis on action, flexibility and interrelatedness, seem well-suited to help state governments adjust to the challenges they face as emerging interventionists in the global economy.

States taking strategic action

As a result, many policy-makers have developed and implemented strategic approaches for state economic development. For example, Texas, Minnesota, Indiana, Maine, California, Pennsylvania, Iowa, Wisconsin and Florida have developed strategies enabling them to be active participants in the new business and economic environment. These states appear to share an understanding that success depends more on their proficiency to create business and employment opportunities within the world economic environment than to compete for existing companies and jobs.

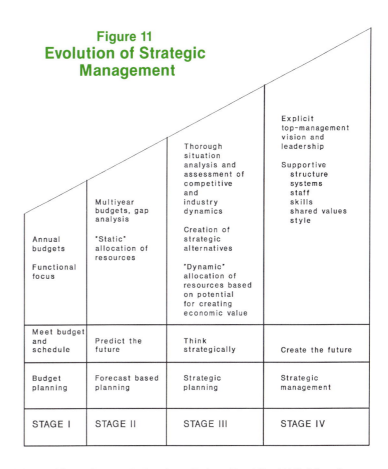

Figure 11
Evolution of Strategic Management

Annual budgets Functional focus	Multiyear budgets, gap analysis "Static" allocation of resources	Thorough situation analysis and assessment of competitive and industry dynamics Creation of strategic alternatives "Dynamic" allocation of resources based on potential for creating economic value	Explicit top-management vision and leadership Supportive structure systems staff skills shared values style
Meet budget and schedule	Predict the future	Think strategically	Create the future
Budget planning	Forecast based planning	Strategic planning	Strategic management
STAGE I	STAGE II	STAGE III	STAGE IV

Adapted from James R. Gardner, Robert Rachlin, H.W. Allen Sweeney, Handbook of Strategic Planning (1986).

The Texas Strategic Economic Policy Commission's strategic plan is designed to improve conditions for economic growth and meet the challenges of an increasingly competitive world market. According to the commission, strategic planning provides a rational basis for making choices, building on the state's strengths and mitigating its weaknesses.

The commission acknowledges that strategic planning by state government is different from that conducted by private firms. For example, the Texas commission advises that the state can influence its economic environment through taxation policies and regulations; policy decisions regarding business activity; quality of life issues such as parks, the environment and crime rates; support to infrastructure by improving highways, airports and water systems; and services such as the quality of public education.

Figure 12
Graphic Rendition of Indiana's Strategic Economic Development Plan

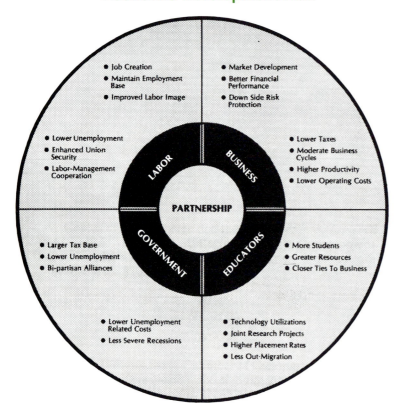

Source: In Step with the Future: Indiana's Strategic Economic Development Plan (1983).

In 1987, the Governor's Commission on the Economic Future of Minnesota presented its initial response to the challenge of the "new global reality," as a blueprint for action by state government in consonance with local government and the private sector. Although the plan is aimed at state government, the commission also recommended broad-based changes in industry, education and the non-profit community.

Minnesota's blueprint envisions the state expanding its links to the global economy. For example, the high technology products that are an important element in the state's recent growth have the highest export share of Minnesota's manufacturing industries. The state's farmers, who always exported significant shares of their products, now are finding the prices for those commodities are set by international market conditions. To measure up to international standards, Minnesota is focusing on technology, superior education and entrepreneurship, all of which crystallize within the state's comprehensive economic development and job creation programs.

Indiana's Strategic Economic Development Plan, developed in 1983, calls for business, government, labor and educators to form a partnership and collaborate in carrying out common economic strategies (see Figure 12). Partnership, as described in the plan, requires all public and private economic development groups to define roles and work together.

Looking Forward: The Update of Indiana's Strategic Economic Development Plan, published in 1988, presents new strategies to guide Indiana's development efforts into the 1990s. It acknowledges that as products from all over the world compete in domestic markets, Indiana businesses that cannot compete in an international market will not survive in the domestic market. With that in mind, the plan proposes new efforts toward working together across sectors and institutions. One example is "boundary spanning" or the building of relationships and alliances among business, government, education and labor.

Maine's Economic Development Strategy, published in 1987, offers similar proposals. So that the state can withstand, and profit from change, the strategy calls for Maine to build a portfolio of assets — one that reflects regional needs and opportunities, balances economic aspirations with the concern for quality of life and distributes the responsibilities, rewards and risks of economic development across all levels of government and all sectors of the economy.

Maine's strategy also includes the building of new partnerships and a greater awareness of market forces. For example, long-standing social issues, such as day care, have become economic issues as well. And education, traditionally the province of local government, now requires new joint action by the public and private sectors to bridge the widening gap between classroom and workplace. In addition, market forces and new technologies require the cultivation of new ideas, close communication with the marketplace and a highly flexible production process and work force.

Ultimately, the strategy is designed to help prepare Maine for change and an unprecedented and accelerating transformation of Maine's economy. The idea is to anticipate the consequences, acquire new skills demanded by the marketplace and assist businesses in exploring the role of new technologies in order to retain their competitive edge. Maine's strategy, simply put, is to preserve its sources of economic strength in an expanding and diversifying economy by taking positive action.

California's strategy for economic development has been in effect since 1983. Since that time, the state has undertaken a variety of efforts to improve its business climate, including reforming its unitary method of taxation, eliminating the inventory tax, reducing capital gains taxes and liberalizing tax loss provisions. But a major feature of California's approach is its promotion of strong partnerships between state government and the private sector. For example, industry and academia are joining forces and pooling resources at the University of California, where a partnership between industry and the state finances microelectronics research.

Of course, in addition to a reordering of its public priorities, California's unique geographic position has had tremendous impact on the state's economy. California, whose major trading partners are Pacific Rim nations, comprises the world's seventh largest economy, and cooperative agreements between the state and 15 leading Japanese banks and trusts link it to nearly 85 percent of all industrial companies in Japan.

Pennsylvania's Economic Development Partnership strategy is based on the belief that a statewide plan is necessary for Pennsylvania to achieve its economic development objectives. The partnership strategy calls on the key sectors of the state — business, labor, education and government — to work together toward common goals. Recognizing that the private sector drives the economy, the role of government is seen as that of catalyst. Accordingly, the most important function of state government is to make certain that the basic foundations — highways, schools, water systems, etc. — are developed and maintained. The strategic approach relies on the governor and the state's entire executive branch to take an active and direct role in the execution of the plan.

The strategy envisions a Pennsylvania economy that moves toward economic diversification through concerted actions by public and private sector leaders and institutions. The state's major economic sectors ideally will approximate the same diversity as the national economy, avoiding narrow concentrations of resources and activities. One approach to that task is joint investment by the commonwealth and the private sector in special institutes and centers to promote research and education which correspond to the needs of targeted Pennsylvania industries. These "magnet facilities" are designed to offer unique research capacities to attract private sector investment and long-term employment growth — all of which contribute to the strategic vision of Pennsylvania as a national and international industrial leader by capitalizing on emerging opportunities in the global marketplace.

In 1987, Iowa published its *Strategic Plan for Economic Growth* in response to what it saw as the "current dynamics of economic change." Although the strategic plan is a set of policy directions to guide the state's economic development actions, it also allows a degree of freedom for politically-realistic implementation. Several actors were involved in its preparation, including the Iowa Department of Economic Development, labor and business representatives, a variety of government agencies and a special research team from the state's universities.

The dynamics of economic change identified in Iowa's strategic plan include the transformation of the United States into a service- and information-based economy, one requiring a highly educated workforce. In response to that transformation, Iowa state government is altering its attitude and emphasis from sole concentration on industrial development to economic development.

Since Iowa's agricultural base is substantially influenced by world markets, the plan advises Iowans to consider seriously the impact the global economy has on their state. World competition requires that Iowa not only develop strategies to strengthen and diversify its economic position, but also recognize and take advantage of the new opportunities that arena affords. In response, Iowa's strategic plan proposes to meet global challenges and competition by building partnerships between the state's public and private leaders and implementing the plan to ensure Iowa's successful participation in the global economy.

The Wisconsin Strategic Development Commission was established in 1984 in an effort to develop that state's first strategic planning initiative for economic development. In its final report, the commission identified and prioritized the state's significant economic development issues and recommended the creation of a state council to carry on its work. Subsequently, the Governor's Advisory Committee on Business Incentives was established in 1987 to explore and suggest guidelines for state incentives to business (see Guidelines for Awarding Business Incentives, p. 21). And in 1988, the Strategic Planning Council was created to continue the commission's effort toward long-range strategic planning for economic development in Wisconsin.

In Florida, collaboration between that state's Chamber of Commerce, Economic Growth and International Development Commission and Department of Commerce resulted in the completion of "Project Cornerstone" in April 1989. The project is a fundamental reassessment of Florida's economic development goals,

strategies and policies through the year 2000.

"Project Cornerstone" gives state and local governments concrete suggestions for modifying law and policies to provide a more "attractive setting for economic development." It identifies high-growth industrial sectors, assesses the likely effectiveness of economic development programs, summarizes the state's economic trends and driving forces and evaluates Florida's strengths and weaknesses. In addition to suggesting ways that economic development organizations and programs might be improved, the project gives special emphasis to how all organizations concerned might better complement each other's efforts.

These and other states — notably Hawaii, Illinois, Rhode Island, South Dakota, Nebraska, Washington, Ohio and Michigan — are using strategic approaches to economic development as a means of addressing major transformations in the global market and determining their destinies in the world economy. More than a management technique, their use of state strategic economic development is an effort toward action-integrated public policy-making. Yet, despite the seemliness of strategic economic development, it is too early to know whether it will be a lasting and reliable approach. However, the quickly-changing global economy with which state governments are challenged is likely to be an unforgiving testing ground for ineffective approaches to state economic development.

Changing Roles in the New Economic Environment

As states have intervened to address the needs of their own economies, the scope of their efforts also has expanded to include more diverse economic development activities, a greater degree of political collaboration and broader public-private cooperation. State economic development activities now require more information and a more thorough working knowledge of business and industrial decision-making, state-of-the-art technology, world market dynamics, entrepreneurship, wealth creation, research and development potential, capital markets, regional economic trends and international trade and investment.

Guidelines for Awarding Business Incentives

In Wisconsin, the following criteria are considered when a decision is made on whether to award business incentives for an economic development project. While a project would not need to do well on each guideline to secure financing, part of the competitive incentive award process would dictate that those projects which best meet the criteria listed below should receive first consideration when funding decisions are made.

State Cost Per Job Created/Retained. The ratio of state dollars requested to the number of jobs created/retained by the project should be calculated for a given period of time. The lower the state cost per job created/retained, the stronger candidate the project is for state funding. This guideline is intended to maximize the potential job creation/retention impact of state funds.

Ratio of Wages to State Cost. The ratio of the annual wages resulting from the project to the state dollars requested should be calculated for a given period of time. The larger the wages created/retained per state dollar, the stronger candidate the project is for state funding, and the sooner the state is likely to recapture its investment through tax revenues. When calculating this ratio, consideration should be given to the various standard wage-levels around the state.

Ratio of Capital Investment to State Cost. The ratio of capital (plant and equipment) investment to the state dollars requested should be calculated. The higher the ratio, the stronger candidate the project is for state funding. Capital investment is a valuable measure in two ways. First, insomuch as the capital investment results in property improvements, it may also result in increased property tax revenues. Second, capital investment is one measure of a company's commitment to remain in the state. The larger a company's investment in plant and equipment, the more likely it is to remain in the state for a long period of time.

The Extent to Which the Business Increases Wealth in the State. The extent to which the business exports goods or services outside state borders should be measured. The greater the

firm's export plans and potential, the stronger candidate the project is for state funding. Businesses which are likely to capture the state's markets which would otherwise be filled by out-of-state firms would also receive special consideration under this guideline. Those businesses which import dollars from out-of-state or which retain dollars in state will increase wealth in the state.

Community Distress. The unemployment and personal income levels of the area in which the project is located should be determined. The higher the unemployment rate or the lower the personal income level, the stronger candidate the project is for state funding. These projects should help reduce unemployment compensation and public welfare costs.

Benefit-Cost Ratio. The likely economic benefits of the project should be compared with the likely public costs of the project. The higher the ratio of the benefits to the costs, the stronger candidate the project is for state funding. The benefits must significantly exceed the costs for state funding to be considered.

Ratio of State Dollars to Private Dollars. The ratio of the state dollars requested to the private dollars committed to the project should be calculated. The larger the private sector investment, the stronger candidate the project is for state funding. Private sector investment is a useful measure for at least three reasons. First, the state would be sharing the risks associated with the project. Second, the private investment means that someone other than the state will be evaluating the risks of the project. Third, insomuch as the business receiving the funds is making a substantial investment in the project, the greater the commitment of the business is likely to be.

Wisconsin Linkages and Multiplier Effect. The extent to which the project is likely to contribute to the growth of existing in-state businesses or is likely to spur the creation of new in-state businesses should be determined. The greater its business development impact, the stronger candidate the project is for state funding. By contributing to such a project, the state

Continued on page 22

A greater degree of political collaboration among public officials, particularly at the state and local level, has become fundamental to the states' new economic role. Once again, the shrinking federal domestic role has been a major factor contributing to this trend. Reductions in federally-funded programs have created the need for stronger relationships between state and local governments — relationships that have taken the form of increased cooperation and information sharing between governors, legislators and mayors. Just as important, new relationships between state government agencies have developed to meet decision-makers' needs to integrate related functional areas — such as education, transportation and the environment — into economic development policy-making. These agencies also are increasingly accountable to a variety of service populations and key stakeholders who are affected by or who can affect economic development — developers, bankers, chambers of commerce, actual or potential employers, neighborhood groups and environmentalists.

The economic environment that has emerged in the 1980s limits state governments' capacity to pursue economic development in older, conventional ways (see Economic Development, p. 23). Increasing national and international competition has hampered industrial recruitment strategies and altered traditional assumptions of what is required to stimulate private investment. However, competition among the states, in itself, is not necessarily unhealthy, and may, in fact, lead to innovation, increased productivity and more creative management of resources. Increasingly, states are inclined to compete over the quality of education, infrastructure and labor force productivity, rather than their tax and financial incentives, which seem to have been employed as defensive measures against regional competition.

States are beginning to realize that competition for scarce industrial prospects creates adversarial relationships that may hinder future economic development. This new intergovernmental, intersectoral arena is compelling states to adopt collaborative strategies, since they can no longer deal with economic problems independently. In this environment, policy-making is dispersed and shared by a multitude of public officials and private sector representatives.

However, as some state governments launch innovative programs to address new economic realities, many feel overwhelmed by the complexities of corporate finance, major technological change, structural shifts in the national economy and the international marketplace. As a

Guidelines . . .
from page 21

would be indirectly aiding other in-state businesses as well, thereby extending the impact of the state incentive.

Protection of the State's Interests. The state must be careful to protect its interests when considering an incentive award. If an incentive package would cause the state budget balance to drop below its statutory minimum, or would jeopardize the state's bond rating, it should not be offered. The financial condition of the company should be examined so that the state does not lose money on a company likely to fail even with the incentive, or so the incentive does not give a company an advantage over its in-state competitors. Also, the state should include a "clawback" provision in all incentive contracts. This provision would enable the state to re-

capture part or all of its financial support should the business fail to meet its obligations under its incentive award contract.

The Extent to Which the Project Builds on Existing Strengths and Resources. The extent to which the project builds on the state's existing strengths and resources should be examined. The greater the degree to which the project builds on existing strengths, the stronger candidate the project is for state funding. Business development projects that would link up with and build on these strengths would add to the state's economic foundation.

Source: Wisconsin Department of Development, Final Report — Governor's Advisory Committee on Business Incentives (1987).

result, active collaboration and information sharing with the private sector has become a significant part of state economic development policy. Public and private sectors can no longer operate in isolation from one another, and increasingly, each sector depends on the vitality and integrity of the other.

For example, increased interdependence is forcing businesses to expand their attention from a known number of shareholders to unknown numbers of stakeholders, including individuals, organized groups, communities and government institutions which, more and more, influence the behavior of businesses. Unilateral action in the private sector has become much more difficult with the active involvement of this expanding network of stakeholders. As a result, firms are rethinking their market positions and competitive strategies, while industries are being restructured.

Disruptions during the past decade, including global competition, technological change, deregulation, volatile money markets and information explosions, have caused major changes in macro-economic and socio-political systems. Destabilization and fluidity, once undesirable, are now the norm in world business. From the perspectives of both businesses and state governments, the shift from national to international competition has created major challenges to survival in a global economy.

State Economic Development Policy Forecast

The tremendous changes that created economic restructuring during the 1980s, will accelerate in the 1990s. State governments will need to be much more flexible, responsive and sophisticated in order to even minimally serve their constituents, and state economic development policy will have to be adjusted to address these realities. Communication and collaboration between the public and private sectors, and among the various levels of government, will need to be far more developed than they currently are.

Successful state economic development efforts will depend on policies that recognize the natural interpolitical and intergovernmental conflicts in the policy-making process and build in ways to resolve or reduce them. Those economic development policies must be linked to the educational, social, cultural and political policies that also affect each state's economic environment. How well states meet these challenges will determine, in large

---

Economic Development

Outdated Strategies	New Strategies
Industrial attraction and plant relocation	Local, "homegrown" business enterprise development
Reliance on federal policy guidance and financial assistance	Reliance on state and local leadership
Focusing on large manufacturing firms	Focusing on smaller and younger firms
Providing low-cost labor	Providing skilled and flexible labor
Providing low-cost land and tax subsidies	Providing accessibility to advanced technology and financial capital
Expansion into regional and national markets	Expansion into international, global markets
Increasing jobs and employment opportunities	Wealth creation and increasing the number of employers

Source: Jeffrey S. Luke, Curtis Ventriss, B.J. Reed, Christine M. Reed, Managing Economic Development: A Guide to State and Local Leadership Strategies (1988).

part, the success of the national economy. In many ways, the policies state governments use to approach their new economic circumstance will provide the basis for national economic policy and economic conditions in the 1990s.

States will need to develop economic development policies that make them competitive participants in the world market, not just competitors with other states. In fact, a state may find that using other states as points of comparison may be misleading, unless its position and potential in the global economy also are heavily weighed. Such a scenario supports the characterization, often promoted by California and Florida, of nation-states that aggressively pursue active and autonomous roles as international business brokers. In such an environment, states will need to develop their economic capacity to create wealth, and their ability to produce valuable goods and services — in both the public and private sectors — rather than just produce jobs. In the rapidly changing world marketplace of the 1990s, jobs will be created, eliminated and altered continuously. But the employment potential for each state will be determined by its economic capacity and the wealth it has created.

In other words, in order for states to generate businesses and jobs which serve the public good in a cost-effective way, they must first enhance their economic capacity and create wealth by designing and executing economic development policies tailored to their own economies. In order to do so, policy-makers will need to evaluate, realistically and comprehensively, the strengths and weaknesses of their own state's economy. Changes in the characteristics of the labor force, the availability of capital resources, the potential for entrepreneurial activity, the productivity of local industry and the effectiveness and equity of the regulatory environment are all factors that will require serious consideration in the policy-making process.

Such policy-making will require increased collaboration and better communication among state government, local government, business and industry. The policies that emerge should create state economies that are evaluated in terms of legislative environments and state government action, as opposed to rankings in business climate studies. Collaborative strategies such as these will recognize and account for the ever-expanding economic development arena in which policies are inextricably linked to almost every other area of state and local government, and where policy-making is shared by a full spectrum of public officials, business and industry representatives and citizen stakeholders.

State policy-makers of the 1990s will continue to face hyper-evolutions in the world economy in which they currently compete. Policy-making and policy-execution necessarily will require approaches that are comprehensive, collaborative, future-oriented and pro-active. The approaches emerging from the 1980s that seem most appropriate for these challenges are variations on private sector strategic planning, management and action techniques. The public sector application of these techniques by at least 17 states has crystallized into what may be appropriately characterized as state strategic economic development.

State Strategic Economic Development

Ideally, state strategic economic development will promote a reconceptualization of economic development, one that initiates comprehensive, strategic policy action. Such action will require an accurate understanding of the issues, the formulation of strategic goals and visions, a formal structure to coordinate and execute initiatives and a continuing and concurrent evaluation process to keep economic development policies in tune with the economic, political and social changes likely to occur in the state's future. Strategic economic development has the potential to create, rather than restrict, the range of future choices by integrating and directing the energies of public, private and non-profit networks, citizen participation and the state's resources toward the development of economic capacity and predictable long-term growth.

This approach appears especially suited to state governments' economic development roles in the global economy of the 1990s (see Figure 13). It recognizes economic, political and social changes that affect a state's economy by constructing policies that address those forces. It reconsiders existing policies and programs in order to determine their effectiveness and it provides a forum for intergovernmental and intersectoral linkage and collaboration.

However, strategic approaches developed in the private sector should be applied carefully and judiciously to public purposes. State governments face problems and predicaments that are quite different from those confronting business and industry. Their goals and operations are drastically different. Different strategies are required for the pursuit of business profit and the provision of public service — strategies determined largely by what is permitted either in the marketplace or in the public regulatory environment. The marketplace is composed of relatively simple circumstances compared to the labyrinth of state government and politics. State governments practicing strategic economic development will have to factor in taxing, regulatory and enforcement responsibilities, as well as expenditure considerations, such as cost-effectiveness and constituent benefits. Likewise, the electoral process, political campaigns, transient leadership, advocacy groups and judicial renderings, as well as the media will require an attentiveness and sensitivity if policies are going to be realistic and have a reasonable chance of success.

There is no single way to approach state strategic economic development. Effective economic development strategies must be crafted with each state's individual economic, political and social circumstances in mind. Regardless of the variation in approach, these strategies should consider each state's internal capabilities as they relate to the dynamics of external change. For economic development, the value of the approach is not in the mechanics of strategic techniques. It is instead in the philosophy of the approach, as a way of thinking and doing, which results in ongoing strategic policy action.

With that in mind, the following policy recommendations are offered as points of reference for state government policymakers contemplating strategic policy action as an approach to economic development in their state.

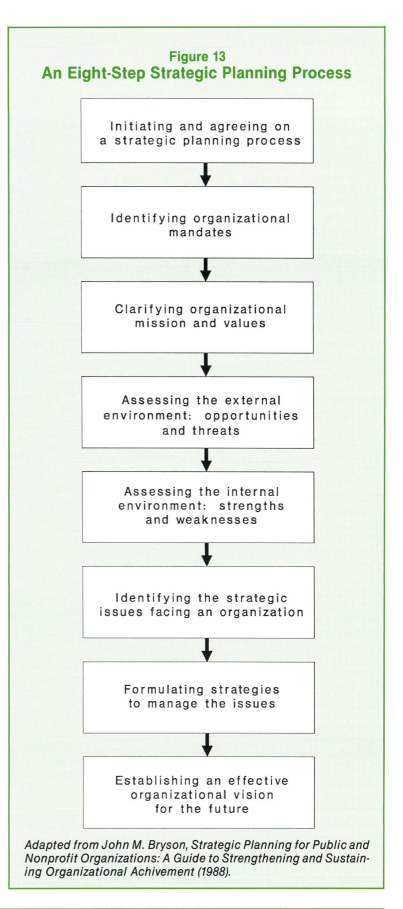

Figure 13
An Eight-Step Strategic Planning Process

Initiating and agreeing on a strategic planning process

Identifying organizational mandates

Clarifying organizational mission and values

Assessing the external environment: opportunities and threats

Assessing the internal environment: strengths and weaknesses

Identifying the strategic issues facing an organization

Formulating strategies to manage the issues

Establishing an effective organizational vision for the future

Adapted from John M. Bryson, Strategic Planning for Public and Nonprofit Organizations: A Guide to Strengthening and Sustaining Organizational Achievement (1988).

Policy Recommendations for Strategic Economic Development

- **Think and Act Strategically Regarding State Economic Development.**

Strategic thinking and action should enable a state to proceed intelligently and with foresight toward its economic development activities. The strategic approach should enhance the state's future position in a global economy of constantly changing markets. It should reflect the state's regional and local needs and opportunities, balance economic goals with quality of life priorities and involve all levels of government and all sectors of the economy. And most important, it should be action-oriented.

- **Create a Stable and Equitable System of State Fiscal, Legal and Regulatory Policies.**

The state's fiscal, legal and regulatory policies create the environment in which the state's businesses prosper or fail. These policies also determine the effectiveness with which state government agencies, local governments, and the private sector are able to interact and coordinate their efforts. These policies should be stable and equitable, both for a better business environment and for a more responsible performance of state services.

- **Develop State Programs and Procedures that are Flexible and Change-Oriented.**

State economic development programs and procedures should be highly flexible in order to survive the turbulence of constant change and rapid innovation that will characterize the global economy of the 1990s. The structure and content of these programs and procedures should enable state governments and businesses to react quickly to market shifts, new technologies and innovative ideas. State governments must prepare to operate in an economic environment in which accelerated change is constant.

- **Prepare to Participate, as a Competitive State, in the Global Economy.**

As a matter of predicament, each state is a participant in a new global economy. In order to prosper, or even survive in the world marketplace, each state must act with initiative and intelligence based on its knowledge of foreign markets, international economics and how its own resources relate to those forces. Each state must know its own strengths and weaknesses, in economic terms, and create policies accordingly.

- **Collaborate with Business and Local Government and Involve People in the State's Economic Destiny.**

State economic development policy-making should be broad-based yet articulate and sensitive to the needs, priorities and expectations of all concerned. The application of emerging policies should involve those same individuals, groups and institutions as a means of ensuring their effectiveness in terms of success and ongoing revision. State governments must be prepared to collaborate with the appropriate political and economic forces operating within their own boundaries if they are to prosper in the 1990s.

- **Create an Efficient State Communications Infrastructure in which Businesses, Local Governments, Schools and Citizen Groups Participate.**

Successful state economic development in the 1990s will require an increasingly sophisticated communications capacity, with regard to both data creation and information exchange. State governments, businesses, schools and other key organizations must collaborate in order to provide information, to interface with one another and to upgrade their systems and data bases. The resulting communications infrastructure should be dynamic and interface effectively with world-wide information-exchange networks.

- **Create Knowledge Regarding State Resources, Business Environments and Economic Development Policies.**

State governments and business and industry need current, complete and comparable information regarding state resources, business environments and economic development policies. Information regarding tax structures, labor, infrastructure, capital markets, business assistance, educational systems, social services, housing, energy and communications must be created and regenerated consistently. The resulting data bases and information systems should be integrated continuously into the states' communication infrastructure.

- **Promote Responsible and Participatory Corporate Citizenship in the State.**

The private sector will play an increas-

ingly important role in state governments' future economic development activities. But if input from the private sector regarding state government policies and regulations is going to be accepted and taken seriously, there must be a growing public-mindedness on the part of business and industry and the emergence of good corporate citizens. Responsible corporate citizenship is an essential factor in the strategic policy-making process.

- **Commit Funding and Support to the State's Economic Mission.**

State strategic economic development is a long-term, broad-based commitment that must be funded and supported on a continuing basis if it is to be successful. State policy-makers in both the public and private sectors must collaborate in an ongoing way in order to provide the stable and reliable base of funding and support that is necessary for successful state economic development in the global arena.

- **Create and Regenerate Wealth and Economic Capacity in the State.**

State strategic economic development should be directed toward the continuing creation of new wealth, in terms of valuable goods and services and the development of the economic capacity to retain and expand the wealth that they have created. Wealth and economic capacity are fundamental to a stable and prosperous state economy. Jobs and businesses are generated from wealth and economic capacity and cannot be long sustained without them.

Appendix A

Methods states use to assess costs and benefits associated with packaging tax and financial incentives for businesses wishing to locate or expand in that state.

Alabama Generally, the Alabama Development Office (ADO) prepares state business comparisons (e.g., Alabama vs. Georgia for industry desiring to locate or expand in Alabama). On a case-by-case basis, the ADO also prepares a tax benefits package for the industry in question. ADO does not presently use any particular method(s) to assess the costs and benefits associated with what the state of Alabama has to offer industry from a tax or financial incentive.

Alaska In preliminary stage of development.

Arizona Different methods: value judgment; input/output model; econometric model.

Arkansas No formal method established.

California Any cost/benefit decision is made during the legislative process if state monies are involved. The most recent examples are the state's Enterprise Zone Program and unitary tax reform.

Colorado Limited benefit analysis now; developing a state development impact model to make public investment/benefit assessments. Job training assistance is assessed on number of trainees and an evaluation of income/taxes paid prior to and after training. Public payback periods are then developed.

Connecticut None.

Delaware 1. Most programs require investment to obtain incentive. 2. Relate paybacks of incentives with added or new personal income tax and corporate taxes (jobs and related personal income tax revenue, income tax and corporate income tax, property tax and gross receipts tax).

Florida Packaged state tax and financial incentives, other than the Economic Development Transportation Fund, take the form of special appropriations by the legislature and must withstand the scrutiny of that procedure.

Georgia Case-by-case evaluation of project and location.

Hawaii No targeted incentives.

Idaho None.

Illinois REMI model used to gauge economic effect. More specific analysis has been developed for foreign investments.

Indiana The state assesses the costs of tax and financial incentives and weighs those against the benefits measured through specific criteria. In some instances, a detailed, project-specific analysis is performed by economists in the Department of Commerce's Division of Economic Analysis to determine the benefits the state will receive in the future as a direct result of the project (both increased revenues and decreased costs, including social costs, are taken into account). This allows a "payback" or "return on investment" analysis of the state's investment in the project. The criteria attempts to account for the value of the project to the state, while paying particular attention to the significance of any public assistance to the private sector investor. Criteria include: extent to which business is or will be a part of the state's economic base; impact of the project on employment and payroll; new capital investment; impact on the affected community; significance of public assistance to the private sector investor.

Iowa Business analysis checklists. Necessary and appropriate review.

Kansas NR

Kentucky All business projects using state financing are evaluated for economic impacts, state and local tax yields and other performance measures. Some larger projects have contractual agreements with the businesses guaranteeing certain levels of employment or other activities (subject to payments for shortfalls or forfeiture of assets).

Louisiana	Customized tax comparison analysis.
Maine	No formal benefit-cost analyses performed.
Maryland	The principal method of packaging both tax and financial incentives is a state enterprise zone: job creation tax credits, real property tax abatements and increased ceilings on business loan guarantees and on grants and loans to the political jurisdictions. Case-by-case basis.
Massachusetts	None.
Michigan	Computer-based estimates of expected tax revenues. Standard methods to assess financial projections and risk on loans.
Minnesota	Estimate taxes created by investment in large projects using a regional input/output model (REMI). On large packages, use an economic model (REMI) to generate state revenue impacts of a business location expansion and compare revenues with state incentives; model not used for smaller community-based products.
Mississippi	No precise method. The state generally relies on broad analyses.
Missouri	Projects analyzed individually.
Montana	The state board of investments uses formulas based on national and state data to evaluate the public costs and benefits of development finance loans and guarantees. The state Department of Revenue is undertaking a study of the costs, benefits and recipients of tax credits and incentives.
Nebraska	Track progress on projects approved for tax credits, focusing on new employment, investment and tax credits received. One particular tracking is the stream of tax revenues in relation to tax credits.
Nevada	None.
New Hampshire	State does not package state tax and financial incentives; offers industrial revenue and mortgage bonds to eligible firms.
New Jersey	None.
New Mexico	NR
New York	Judgments made on a case-by-case basis regarding level of incentives that can be justified by the ultimate benefits of the project.
North Carolina	All considerations are project-specific.
North Dakota	State receives testimony from public hearings at local level to determine if there is unfair competition.
Ohio	NR
Oklahoma	Impact analysis of proposed industry investment versus tax and financial incentives provided.
Oregon	State determines tax and financial incentives granted to new and expanding businesses according to the number of family wage jobs created. There is a correlation between the quantity of incentives granted and the amount of personal income tax dollars generated through the creation of new jobs.
Pennsylvania	The single point of responsibility for this function is the Governor's Response Team, which has the authority to arrange commitments across departments. It uses policy targets derived from the state strategy, plus reasonable administrative guides. Assessment is viewed as an art, not a science. Rigid, purely ''quantifiable'' criteria have serious limitations in dealing with the variety of real situations and in accounting for perceived political/psychological impact.
Rhode Island	Evaluates job creation, income received and tax revenue gained by local community and state.
South Carolina	Economic impact analysis.
South Dakota	The state Department of Labor's Labor Market Information Center reviews each loan application to the state's low interest loan program for the firm's potential impact on the community.
Tennessee	NR
Texas	The state Comptroller's Office uses models for determining costs and benefits.
Utah	Not yet developed.

Vermont Considers community needs, ability to meet employment levels, wage rates, non-polluters and investment to be made in the state.

Virginia Virginia's overall tax and financial incentives are limited and discretionary incentives for individual companies are extremely limited. At the state level, the main incentives are industrial training, industrial access road funds and rail access funds. Small businesses may be eligible for the Virginia Small Business Financing Authority's taxable financing and working capital loan guarantee programs.

Washington Program-by-program basis. Although analysis varies with the program, it tends to focus on dollars per job, tax return to state, return on investment, preference toward incentives for distressed areas.

West Virginia NR

Wisconsin Guidelines (refer to "Guidelines for Awarding Business Incentives," p. 21).

Wyoming NR

NR = No Response to survey question.

Source: The Council of State Governments, "Business Tax and Financial Incentives: A Survey of State Economic Development Agencies (fifty-state survey)," summer/fall, 1988.

Appendix B

Information and technical assistance needed by states in business recruitment and retention efforts.

Alabama New and innovative programs that might assist industrial prospects.

Alaska Computerized (online) world-wide market data; world-wide raw material data (market price, availability, etc.).

Arizona Computer models, software, training on evaluation models; information about our competitors (i.e., other states).

Arkansas National guidelines to determine economic impact of new or expanding plants in terms of new tax revenues generated for local and state governments, new businesses for vendors (primarily small businesses), value of construction, comparable data. U.S. Chamber of Commerce study on impact of 100 new jobs is currently used, but that is too general for close evaluation.

California NR

Colorado A general payback model for all states using incentives. Information on other state incentive programs.

Connecticut More specific information on industries; more responsive network of services and incentives packages which we could offer to businesses.

Delaware Long-term goals and objectives of the team; realistic financial outlook; competitive position in their industry; comprehensive nationwide tax and incentive comparison base.

Florida Accurate, current, detailed wage data and detailed, current information about local markets.

Georgia None.

Hawaii Comprehensive document listing all state incentives, programs and services as an economic development tool.

Idaho Data on the cost of doing business in every state (from an unbiased source), including, but not limited to:

taxes (in comparable format), shipping costs, detailed wage rates, energy costs, and utility costs. Information currently disseminated by many states is either inaccurate, incomplete, biased, not in a form comparable with that from other states or all of these. Better economic information on what types of businesses are prospering in particular geographic areas and analysis of the economic environment in which they operate (shift-share analysis might be a part of this).

Illinois "We finally find that we are more knowledgeable about most companies and businesses than they are about our programs. We have the capacity to know if the industry is growing or shrinking. When we encounter a complex technology we seek outside help and find it."

Indiana NR

Iowa Costs of doing business in other states (building costs, labor costs, etc.). Incentives other states are offering specific businesses. Predictability of job estimates and sales estimates of companies.

Kansas Capacity to evaluate the impact of an incoming firm on the state or local economy. Method to determine the benefits or costs of any incentive package that may be offered to an incoming firm.

Kentucky Good five- to 10-year forecasts of the performance of specific industry groups at the state and substate levels. Data on wages, by occupation, for small areas (counties, cities) on a regular basis.

Louisiana Forecasts of growing markets, both nationally and internationally.

Maine NR

Maryland The state Department of Economic and Employment Development has successfully used an economic forecasting model called IMPLAN (run by the U.S. Department of Agriculture, U.S. Forest Service, Ft. Collins, Colorado), which identifies the impacts of a company on the local economy by sales, income, employment and taxes.

Massachusetts Information on manpower training and recruitment.

Michigan Information on other states' incentive packages.

Minnesota Certainty that the investment would not have been made without the incentive. Better targeting information for communities. Market information. Expansion/relocation startups.

Mississippi Reliable market and labor availability data. Mississippi is a rural state with most of its communities having populations of less than 25,000. Also, the state is relatively distant from most of the major market areas. Under these circumstances, the state must be able to pinpoint market and labor availability data as accurately as possible. Unfortunately, the information needed to accomplish this task is not readily available.

Missouri Information regarding target industries identified for potential development. One area that has been identified for improvement is education and job training development.

Montana More information about in-state businesses: their capabilities, markets, plans and problems.

Nebraska Better information on the market "niche" of the firm and greater confidence in, or better information about, the projected costs and revenues of the firm.

Nevada Small business financing.

New Hampshire Information on the types and value of products exported annually, by state, to country of destination. This information would help the state improve its technical assistance to small- and medium-sized firms wishing to export.

New Jersey None.

New Mexico Financial expertise; market advice and analysis.

New York Information about each company's overall cost structure and any locational differentials.

North Carolina Good, reliable comparison of data among states nationwide; current data totally unsatisfactory.

North Dakota Need technology transfer from research institutions.

Ohio None.

Oklahoma NR

Oregon More information describing the incentives and efforts other states are concentrating on in recruiting and retaining businesses.

Pennsylvania "Industry market data is not the central problem. This data is now more available than currently perceived. Negotiations with individual companies often mean that the information you would most want to have is what they are least likely to provide. The best you can do is have a very clear idea at the outset of what you are looking for (i.e., strategy/policy/targets)."

Rhode Island None.

South Carolina NR

South Dakota In order to avoid the "shotgun" approach to recruiting firms, the Governor's Office of Economic Development and the city of Sioux Falls hired outside consultants to form targeted industry studies. Also, within the past two years, Centers for Innovation, Technology and Enterprise (CITEs) have been established on the campuses of state-supported colleges and universities to facilitate the transfer of research to the state's private sector.

Tennessee None.

Texas None.

Utah The true costs and benefits of business growth; identification of incentives that are appropriate (and cost effective) to offer.

Vermont NR

Virginia NR

Washington	Generally, information brokering assistance; what other states are doing, emerging issues; programs, strategies to deal with plant closings; special needs of rural areas, inner city distressed, and innovative approaches to financing needs of small and export businesses.	West Virginia	NR
		Wisconsin	The ability to predict or to uncover potential business closings or relocations.
		Wyoming	NR

NR = No Response to survey question.

Source: The Council of State Governments, "Business Tax and Financial Incentives: A Survey of State Economic Development Agencies (fifty-state survey)," summer/fall, 1988.

**Biggest obstacle to state economic development officials' efforts
to encourage businesses to locate or expand in their states.**

Alabama Perceived image.

Alaska Attitude; lack of knowledge about the state and what it has to offer.

Arizona Lack of incentives; inadequate financial resources for economic development; lack of sufficient funds for advertising and other economic development activities.

Arkansas For new businesses coming in, it is a combination of lack of an image and inflexibility regarding our ability to compete with incentive packages which include no property taxes, free land or buildings, large-scale training programs, quality/productive training programs, and infrastructure funds immediately available to solve shortcomings of community systems. For expansions, the main problem is lack of low-cost capital other than tax-free industrial development bonds, which are sunsetting.

California Cost (perception only since state has many good low-cost areas). The erroneous perception that California cares more about its environment than its business; those two are not mutually exclusive.

Colorado Antiquated business assistance attitudes; state constitutional prohibitions on state assistance to private businesses; restrictive banking regulations.

Connecticut High cost of living.

Delaware Unrealistic and unreasonable demands from clients; size and labor availability; availability of "spec" buildings; environmental issues and concerns; transportation access, especially air.

Florida State is working to improve its educational system at all levels to ensure national competitiveness. Some parts of Florida have a backlog of transportation and other public facilities' needs which discourage business investment. For some firms that have high transportation costs and serve a national market, Florida's geographic location can create a perception of cost disadvantages. However, complete information about beneficial "back-haul" rates, and our proximity to large markets and transportation facilities usually allay these fears.

Georgia Restraints naturally imposed by state bureaucratic system.

Hawaii Serious business image.

Idaho Of those elements that cannot be controlled, location. Of those that can be controlled, perception of the state by those outside of it.

Illinois Lack of funding.

Indiana The perception of Indiana as a state in which the business climate is not attractive. Specifically, the perceptions that union and management conflicts are prone to occur; employee wages are high, and taxes on businesses are high.

Iowa Need exposure.

Kansas State's location in the rural Midwest and a slow-growing population base which translates into a slow-growing work force. In addition, the state's relative distance from the nation's major urban concentrations on the East and West coasts has in many cases proven to be a detriment to attracting industry. (Respondent indicated that outsiders' opinions of the Midwest may create greater disadvantages for Midwestern states than those that actually exist.)

Kentucky Education level of the work force.

Louisiana High business taxes (inequitable tax application); education; image.

Maine A perceived labor shortage due to the state's very low unemployment rate.

Maryland The state does not consider that it has major obstacles which hinder plant location and expansion decisions. However, there are other issues the state recognizes are being perceived as drawbacks, such as the absence of right-to-work legislation and the lack of a venture capital program.

Massachusetts	Labor supply.	**Pennsylvania**	Many obstacles at both macro- and micro-levels. At the most basic level, a lack of consensus among government levels and between sectors as to what the policy targets should be, what causes distress and what is truly effective or not. Periodic, participatory, strategy-building processes help in this regard.
Michigan	Misperceptions about business climate.		
Minnesota	Expansion rate is very high, no serious obstacles. Some possible constraints: available pool of top technical people; perceptions of high taxes; getting a consensus among business, labor and public policy on business climate.		
		Rhode Island	Prior to mid-1980s, state was perceived by the business community as anti-business; in 1985, Rhode Island made sweeping changes in its laws affecting business.
Mississippi	Proximity to markets: consumer, commercial and industrial. Most major market areas are at least 400 miles from the center of the state; i.e., there are few major market areas within Mississippi or its four bordering states.		
		South Carolina	NR
		South Dakota	Awareness, by the rest of the country, of what South Dakota is and has to offer in terms of quality of life and pro-business climate.
Missouri	Rural image.		
Montana	Some problems in capital access.		
Nebraska	Getting the rest of the world to "look seriously" at the state.	**Tennessee**	Advertising budget too low.
Nevada	None.	**Texas**	No tax incentives (i.e., big packages), because state taxes are already low.
New Hampshire	Insuring the availability of skilled workers in industries where new technologies are emerging.	**Utah**	Geographic isolation; relatively small nearby markets; perception of monoculture.
New Jersey	Labor shortage.		
New Mexico	Lack of adequate means of finance; venture capital; lack of amenities in many areas; resistance to economic development by certain elements.	**Vermont**	NR
		Virginia	NR
New York	Misleading perception that New York is a high-tax, high-cost state.	**Washington**	Constitutional prohibition on lending of the state's credit (state and local); state literally interprets this provision; provision limits tools for economic development, has dampening effect on economic creativity.
North Carolina	Long-term: education and training; utilities; highways; and the fact that each industry requirement is different.		
North Dakota	For relocating firms, the image of North Dakota by rest of country. For expanding and retaining firms, access to capital (especially long-term).	**West Virginia**	NR
		Wisconsin	Lack of accurate information to dispel a negative high-tax image or to combat misperception about the state, i.e., state is not well-known throughout the U.S.
Ohio	None.		
Oklahoma	NR		
Oregon	Proximity to major U.S. markets (i.e., the Eastern seaboard, Southern California, the Gulf states).	**Wyoming**	Lack of proximity to major markets, inadequate air transportation network; image problem.

NR = No Response to survey question.

Source: The Council of State Governments, "Business Tax and Financial Incentives: A Survey of State Economic Development Agencies (fifty-state survey)," summer/fall, 1988.

Appendix D

Types of new businesses that would be most beneficial to each state.

Alabama — Emerging industries: bio-sciences, interstate commerce, warehousing, tourism, aerospace and defense-related equipment and research, wholesale and retail trade, services, food processing, electrical and electronic equipment, chemicals.

Alaska — Businesses that can add value to basic resources (mining/minerals, petroleum, timber, fishing).

Arizona — Businesses that pay higher wages, are not high-volume water users and are lower pollution generators.

Arkansas — Any manufacturing operation requiring higher skill levels and a higher percentage of workers with degrees. This is necessary to increase income per capita and employ more college graduates who are now migrating to other states. Arkansas also needs more "home-grown" industry; more than 60 percent is now headquartered in other states.

California — Leading-edge technologies, which in turn expands the manufacturing base.

Colorado — High-technology, micro-biology, aerospace, financial services, light assembly/manufacturing.

Connecticut — Businesses capable of innovating and producing new products and processes; high employment growth cost; high-technology, high value-added firms.

Delaware — High-technology firms that mesh with state's educational strengths (composites); light manufacturing; distribution and warehousing; corporate headquarters.

Florida — The State Plan identifies several beneficial new businesses: corporate or regional headquarters or distribution centers; research and development facilities; international banking and foreign investment; motion picture, television and record production and international trade through Florida's ports. Additional target industries have been identified by the state Department of Commerce: semiconductors; medical instruments; aircraft parts; equipment and maintenance; computer and data processing services; communications equipment and services; instruments and controls; sports-related manufacturing; motion picture services and professional sport activity.

Georgia — Diversify from the visitor industry, e.g., agriculture, manufacturing.

Hawaii — Service-oriented, high-growth, diversification-producing.

Idaho — Value-added industries using the existing resource base (timber, mining, agricultural products). Financial, communications and information-based services. High-tech research and manufacturing facilities.

Illinois — Eight targeted industries. Advertising for three: plastics, food processing and automotive.

Indiana — Businesses that will provide, from their Indiana operations, products or services to be sold in multi-state, national or international markets. In addition, beneficial new businesses would be those that diversify state's industrial base.

Iowa — Insurance, biotechnology, auto parts and accessories, telecommunications, plastic.

Kansas — State would benefit from being able to expand its current support resources, i.e., the attraction of additional food processors to utilize the state's agricultural products. In addition, firms needing the support services that are currently available to serve Kansas' major aviation industry would find the state has a large high-technology work force and high-technology support services of all types.

Kentucky — Manufacturing, entrepreneurial activities in most industries, agribusiness, facilities and services for tourism.

Louisiana — Food processing/distribution, refined petrochemical products, most secondary manufacturing.

Maine	Any business providing quality jobs without adverse environmental or other consequences.	**New York**	Firms in growth industries with high local multipliers, i.e., industries that would use locally-produced raw materials, components, supplies, subassemblies, etc.
Maryland	Basic manufacturing and service industries, especially medical products; metalworking; printing; high technology and international firms; research and development firms, especially biotechnology; and corporate and regional headquarters.	**North Carolina**	Same types the state is currently locating.
		North Dakota	Aerospace, biotechnology (biogenetic engineering), metal fabrication, computer software, remote sites for service industry, garment industry, apparel industry, wood products, food processing.
Massachusetts	Centers of excellence programs; employing technology in basic industry.		
Michigan	Technology-based manufacturing, but see significant benefits in all sectors.	**Ohio**	Manufacturing, industrial businesses, high tech.
Minnesota	Basic industries, building on state's comparative advantages, including natural resource base and technical/skilled labor pool. High technology — medical, computer, instruments; manufacturing — particularly in rural areas; natural resources — wood products, mining, agriculture-related, recreation; graphics — printing, design, media.	**Oklahoma**	NR
		Oregon	Businesses that create stable, family wage level employment opportunities.
		Pennsylvania	New businesses in emerging, high growth industries.
		Rhode Island	Health industry, marine electronics, biotechnology and insurance related business.
Mississippi	Vertically-integrated manufacturing facilities.		
Missouri	Auto suppliers, communications, food processors, office/distribution, electronics, plastics/metals manufacturing.	**South Carolina**	NR
		South Dakota	Examples from targeted studies: medical products, plastic products, electronics, frozen food specialties, miscellaneous aircraft equipment and parts, agriculture-related (to build on the present economy); services — international trend; high tech — looking to the future.
Montana	Distance-independent, high value-added manufacturing and service firms.		
Nebraska	Businesses that have the greatest opportunity for profitability are those that utilize (add value to) agricultural products, plus those that have an advantage in a location central to the nation. Most beneficial to the state's future are businesses that diversify and expand the state's economic base.		
		Tennessee	Transportation; non-electrical machinery; electrical and electronic machinery and equipment; rubber and plastic products; food products; chemicals and allied products; printing and publishing; instruments and related products.
Nevada	Firms requiring well-educated, skilled workers and executives.		
New Hampshire	Small high-tech, high value-added businesses that pay above-average wages.	**Texas**	Small business. Diversify, away from oil and real estate.
		Utah	Businesses that would pay high wages, high taxes and not be detrimental to the environment.
New Jersey	None in particular.		
New Mexico	Businesses processing the agricultural production of the state and particularly, processing the state's mineral resources.	**Vermont**	NR
		Virginia	Wide variety of manufacturing and non-manufacturing facilities.

Washington   Trade-dependent state. Pacific Rim companies that are involved in export activities; service companies; potential for tourism-related businesses; value-added processing.

West Virginia   Targeted industries include food processing, wood processing, distribution, medical-related facilities.

Wisconsin   State has not targeted specific industries. However, export-based industries which bring new dollars into the state (i.e., manufacturing, tourism) are our current focus for development efforts.

Wyoming   Wide variety of manufacturing and non-manufacturing facilities are targeted for each of the state's regions in *Wyoming: A Competitor for Jobs and Growth*.

NR = No Response to survey question.

Source: The Council of State Governments, "Business Tax and Financial Incentives: A Survey of State Economic Development Agencies (fifty-state survey)," summer/fall, 1988.

**Private-sector approaches to strategic planning and their applicability
to the public and nonprofit sectors**

*Adapted from John M. Bryson and William D. Roering, "Applying Private Sector Strategic Planning in the Public
Sector," American Planning Association Journal (Winter 1987), pp. 12-14.*

Approach: Harvard policy model (Andrews, 1980; Christensen et al., 1983)

Key Features	Primarily applicable at the strategic business unit level. SWOT (Strengths, Weaknesses, Opportunities, Threats) analysis. Analysis of management's values and social obligations of the firm. Attempts to develop the best "fit" between a firm and its environment; i.e., best strategy for the firm.	**Strengths**	Systematic assessment of strengths and weaknesses of firm and opportunities and threats facing firm. Attention to management values and social obligations of the firm. Systematic attention to the "fit" between the firm and its environment. Can be used in conjunction with other approaches.
Assumptions	Analysis of SWOTs, management values and social obligations of firm will facilitate identification of the best strategy. Agreement is possible within the top management team responsible for strategy formulation and implementation. Team has the ability to implement its decisions. Implementation of the best strategy will result in improved firm performance (an assumption held in common with all strategic planning approaches).	**Weaknesses**	Does not offer specific advice on how to develop strategies. Fails to consider many existing or potential stakeholder groups.
		Applicability to the public and nonprofit sectors	Organizations: Yes, if a strategic planning unit can be identified and additional stakeholder interests are considered, and if a management team can agree on what should be done and has the ability to implement its decisions. Functions: SWOT analysis is applicable. Communities: SWOT analysis is applicable if what is "inside" and "outside" can be specified.

Approach: Strategic planning systems (Lorange, 1980; Lorange et al., 1986)

Key Features	Systems for formulating and implementing important decisions across levels and functions in an organization. Allocation and control of resources within a strategic framework and through rational decision making. Attempts to comprehensively cover all key decision areas.		Information on performance is available at reasonable cost.
		Strengths	Coordination of strategy formulation and implementation across levels and functions. Can be used in conjunction with other approaches.
Assumptions	Strategy formulation and implementation should be rational and anticipatory. An organization's strategies should form an integrated whole. The organization can control centrally all or most of its internal operations. Goals, objectives, and performance indicators can be specified clearly.	**Weaknesses**	Excessive comprehensiveness, prescription and control can drive out attention to mission, strategy and organizational structure. The information requirements of planning systems can exceed the participants' ability to comprehend the information.

Strategic planning systems — continued

Applicability to the public and nonprofit sectors	Organizations: Less comprehensive and rigorous forms of private sector strategic planning systems are applicable to many public and nonprofit sector organizations.		Functions: Necessary conditions for strategic planning systems to succeed are seldom met. Communities: Unlikely.

Approach: Stakeholder management (Freeman, 1984)

Key Features
Identification of key stakeholders and the criteria they use to judge an organization's performance.
Development of strategies to deal with each stakeholder.

Assumptions
An organization's survival and prosperity depend on the extent to which it satisfies its key stakeholders.
An organization's strategy will be successful only if it meets the needs of key stakeholders.

Strengths
Recognition that many claims, both complementary and competing, are placed on an organization.
Stakeholder analysis (i.e., a listing of key stakeholders and of the criteria they use to judge an organization's performance).
Can be used in conjunction with other approaches.

Weaknesses
Absence of criteria with which to judge different claims.
Need for more advice on how to develop strategies to deal with divergent stakeholder claims.

Applicability to the public and nonprofit sectors
Organizations: Yes, as long as agreement is possible among key decision makers over who the stakeholders are and what the organization's responses to them should be.
Functions: Yes, with the same caveats.
Communities: Yes, with the same caveats.

Approach: Portfolio methods (Henderson, 1979; Wind and Mahajan, 1981; MacMillan, 1983)

Key Features
A corporation's businesses are categorized into groups based on selected dimensions for comparison and development of corporate strategy in relation to each business.
Attempts to balance a corporation's business portfolio to meet corporate strategic objectives.

Assumptions
Aggregate assessment of a corporation's various businesses is important to the corporation's success.
Resources should be channeled into the different businesses to meet the corporation's cash flow and investment needs.
A few key dimensions of strategic importance can be identified against which to judge the performance of individual businesses.
A group exists that can make and implement decisions based on the portfolio analysis.

Strengths
Provides a method for evaluating a set of businesses against dimensions that are deemed to be of strategic importance to the corporation.
Provides a useful way of understanding some of the key economic and financial aspects of corporate strategy.
Can be used as part of a larger strategic planning process.

Weaknesses
Difficult to know what the relevant strategic dimensions are, what the relevant entities to be compared are, and how to classify entities against dimensions.
Unclear how to use the tool as part of a larger strategic planning process.

Applicability to the public and nonprofit sectors
Organizations: Yes, if economic, social, and political dimensions of comparison can be specified, entities to be compared can be identified, and a group exists that can

Portfolio methods — continued

Applicability to the public and nonprofit sectors (continued)	make and implement decisions based on the portfolio analysis. Functions: Yes, with the same caveats. Communities: Yes, with the same caveats.

Approach: Competitive analysis (Porter, 1980, 1985; Harrigan, 1981)

Key Features Analysis of key forces that shape an industry, e.g., relative power of customers, relative power of suppliers, threat of substitute products, threat of new entrants, amount of rivalrous activity, exit barriers to firms in the industry.

Assumptions Predominance of competitive behavior on the part of firms within an industry.
The stronger the forces that shape an industry, the lower the general level of returns in the industry.
The stronger the forces affecting a firm, the lower the profits for the firm.
Analysis of the forces will allow one to identify the best strategy whereby an industry can raise its general level of returns and whereby a firm within an industry can maximize its profits.

Strengths Provides a systematic method of assessing the economic aspects of an industry and the strategic options facing the industry and specific firms within it.
Gives relatively clear prescriptions for strategic action.
Can be used as part of a larger strategic planning process.

Weaknesses Sometimes difficult to identify what the relevant industry is.
Excludes consideration of potentially relevant noneconomic factors.
Tends to ignore the possibility that organizational success may turn on collaboration, not competition.

Applicability to the public and nonprofit sectors Organizations: Yes, for organizations in identifiable industries (e.g., public hospitals, transit companies, recreation facilities) if a competitive analysis is coupled with a consideration of noneconomic factors and if the possibility of collaboration is also considered.
Functions: Yes, if the function equates to an industry.
Communities: No.

Approach: Strategic issues management (Ansoff, 1980; King, 1982; Pflaum and Delmont, 1987)

Key Features Attention to the recognition and resolution of strategic issues.

Assumptions Strategic issues are issues that can have a major influence on the organization and must be managed if the organization is to meet its objectives.
Strategic issues can be identified by the use of a variety of tools (e.g., SWOT analyses and environmental scanning methods).
Early identification of issues will result in more favorable resolution and greater likelihood of enhanced organizational performance.
A group exists that is able to engage
in the process and manage the issue.

Strengths Ability to identify and respond quickly to issues.
Has a "real time" orientation and is compatible with most organizations.
Can be used in conjunction with other approaches.

Weaknesses No specific advice is offered on how to frame issues other than to precede their identification with a situational analysis.

Strategic Issues management — continued

| Applicability to the public and nonprofit sectors | Organizations: Yes, as long as there is a group able to engage in the process and manage the issue. Functions: Yes, with the same caveat. Communities: Yes, with the same caveat. |

Approach: Strategic negotiations (Pettigrew, 1977; Fisher and Ury, 1981; Allison, 1971)

Key Features Bargaining and negotiation among two or more players over the identification and resolution of strategic issues.

Assumptions Organizations are "shared power" settings in which groups must cooperate, bargain, and negotiate with each other in order to achieve their ends and assure organizational survival.
Strategy is created as part of a relatively constant struggle among competing groups in an organization.
Strategy is the emergent product of the partial resolution of organizational issues.

Strengths Recognizes that there are many actors in the strategy formulation and implementation process and that they often do not share common goals.
Recognizes the desirability of bargaining and negotiation in order for groups to achieve their ends and to assure organizational survival.
Can be used in conjunction with other approaches.

Weaknesses Little advice on how to ensure technical workability and democratic responsibility — as opposed to political acceptability — of results.
No assurance that overall organizational goals can or will be achieved; there may not be a whole equal to, let alone greater than, the sum of the parts.

Applicability to the public and nonprofit sectors Organizations: Yes. Functions: Yes. Communities: Yes.

Approach: Logical incrementalism (Quinn, 1980; Lindblom, 1959)

Key Features Emphasizes the importance of small changes as part of developing and implementing organizational strategies.
Fuses strategy formulation and implementation.

Assumptions Strategy is a loosely linked group of decisions that are handled incrementally.
Decentralized decision making is both politically expedient and necessary.
Small, decentralized decisions can help identify and fulfill organizational purposes.

Strengths Ability to handle complexity and change.
Attention to both formal and informal processes.
Political realism.
Emphasis on both minor and major decisions.
Can be used in conjunction with other approaches.

Weaknesses No guarantee that the loosely linked, incremental decisions will add up to fulfillment of overall organizational purposes.

Applicability to the public and nonprofit sectors Organizations: Yes, as long as overall organizational purposes can be identified to provide a framework for incremental decisions. Functions: Yes, with the same caveat. Communities: Yes, with the same caveat.

Approach: Framework for innovation (Taylor, 1984; Pinchot, 1985)

Key Features Emphasis on innovation as a strategy.

Reliance on many elements of the other approaches and specific management practices.

Assumptions Change is unavoidable, and continuous innovation to deal with change is necessary if the organization is to survive and prosper.

A "vision of success" is necessary to provide the organization with a common set of superordinate goals toward which to work.

Innovation as a strategy will not work without an entrepreneurial company culture to support it.

Strengths Allows innovation and entrepreneurship while maintaining central control on key outcomes.

Fosters a commitment to innovation.

Can be used in conjunction with other approaches.

Weaknesses Costly mistakes usually are necessary as part of the process of innovation.

Decentralization and local control result in some loss of accountability.

Applicability to the public and private sectors

Organizations: Yes, but the public is unwilling to allow public organizations to make the mistakes necessary as part of the process, and development of an overall framework within which to innovate and maintain central control over key outcomes is difficult.

Functions: Yes, but with the same caveats.

Communities: Yes, with same caveats.

Appendix F
The Strategic Planning Model

Strategic planning is neither the personal vision of the chief executive officer nor a collection of unrelated plans drawn up by department heads. Strategic planning is done by the top line officers of the organization, from the chief executive officer through the upper levels of middle management. It is not done by planners.

The strategic plan does not substitute numbers for important intangibles, such as human emotions, but it does use computers and quantification to illuminate choices. It attempts to go beyond a simple surrender by the organization to environmental conditions, and in this sense it is by no means a way of eliminating risks. What a strategic plan does is place line decision makers in an active rather than a passive position about the future of their organization. It incorporates an outward-looking, aggressive focus that is sensitive to environmental changes, but does not assume that the organization is necessarily a victim of changes in its task environment.

Strategic planning concentrates on decisions rather than on extensive documentation, analyses, and forecasts. In this sense it attempts to free itself of the constraints alleged by those who criticize rationalist approaches for costing more than they save and for having no particular bearing on decision making in an organization. Because it is decision-oriented, strategic planning blends economic and rational analysis, political values, and the psychology of the participants in the organization. To do this, strategic planning must be highly participatory and tolerant of controversy. This participatory aspect of strategic planning leads strategic planners to concentrate on the fate of the whole organization above all other concerns; the fate of subunits in the organization is clearly secondary to this overriding value.

In these ways, strategic planning is an attempt to reconcile the rationalist and incrementalist approaches to public policy formation.

Source: Nicholas L. Henry, Public Administration and Public Affairs (1986).

The Strategic Planning Model of Public Policy-Making

Incrementalist Resources*

traditions, values, and aspirations of agency and its personnel	budgetary, political, managerial and intellectual resources of agency and its line personnel	agency leadership: abilities and policy priorities

↓ ↓ ↓

PUBLIC SECTOR STRATEGIC PLANNING

↑ ↑ ↑

analyses of long-term environmental trends: threats and opportunities	analyses of short-term political trends: threats, opportunities, perceptions and directions	interagency competition: perceptions and directions

Rationalist Resources**

*Incrementalism considers only a limited selection of policy alternatives that are provided to policy-makers, each alternative representing only an incremental change in the status quo.
**Rationalism considers all value preferences, discovers all policy alternatives, knows the consequences of each alternative, and ultimately selects the most efficient policy.

Adapted from Nicholas L. Henry, Public Administration and Public Affairs (1986).

Bibliography

Books

Ackoff, Russell L.; Broholm, Paul; and Snow, Roberta. *Revitalizing Western Economies*. San Francisco: Jossey-Bass Publishers, 1984.

Ameritrust/SRI. *Indicators of Economic Capacity*. Cleveland, 1986.

Below, Patrick J.; Morrisey, George L.; and Acomb, Betty L. *The Executive Guide to Strategic Planning*. San Francisco: Jossey-Bass Publishers, 1984.

Bryson, John M. *Strategic Planning for Public and Nonprofit Organizations: A Guide to Strengthening and Sustaining Organizational Achievement*. San Francisco: Jossey-Bass Publishers, 1988.

Choate, Pat, and Linger, J.K. *The High-Flex Society: Shaping America's Economic Future*. New York: Alfred A. Knopf, 1986.

Clarke, Marianne K. *Revitalizing State Economies: A Review of State Economic Development Policies and Programs*. Washington, D.C.: Center for Policy Research and Analysis, National Governors' Association, 1986.

Congressional Budget Office. *The Federal Role in State Industrial Development Programs*. Washington, D.C.: The Congress of the United States, 1984.

Corporation for Enterprise Development. *Making the Grade: The Development Report Card for the States*. Mt. Auburn Associates and the Institute on Taxation and the Institute on Taxation and Economic Policy, 1987.

Eisinger, Peter K. *The Rise of the Entrepreneurial State: State and Local Government Development Policy in the United States*. Madison, Wis.: The University of Wisconsin Press, 1988.

Fosler, R. Scott. ed. *The New Economic Role of the American States: Strategies in a Competitive Economy*. New York: Oxford University Press, 1988.

Gardner, James R.; Robert Rachlin; and H.W. Sweeney, editors. *Handbook of Strategic Planning*. New York: John Wiley & Sons, 1986.

Hayden, Catherine L. *The Handbook of Strategic Expertise*. New York: The Free Press, a Division of Macmillan, Inc., 1986.

Henry, Nicholas L. *Public Administration and Public Affairs*. Englewood Cliffs, New Jersey: Prentice-Hall, 1986.

John, DeWitt. *Shifting Responsibilities: Federalism in Economic Development*. Washington, D.C.: National Governors' Association, 1987.

Kline, John M. *State Government Influence in U.S. International Economic Policy*. Lexington, Mass.: D.C. Heath and Company, 1983.

Luke, Jeffrey S.; Ventriss, Curtis; Reed, B.J.; and Reed, Christine M. *Managing Economic Development: A Guide to State and Local Leadership Strategies*. San Francisco: Jossey-Bass Publishers, 1988.

McKenzie, Richard. *The Global Economy and Government Power*. St. Louis: Center for the Study of American Business, Washington University, 1989.

Milward, H. Brinton, and Newman, Heidi Hosbach. *State Incentive Packages and the Industrial Location Decision*. Lexington, Ky.: The Center for Business and Economic Research, University of Kentucky, 1987.

Thomas, Joe G. *Strategic Management: Concepts, Practice and Cases*. New York: Harper & Row Publishers, 1988.

Thornton, Grant. *Ninth Annual Manufacturing Climates Study*. Chicago: Grant Thornton, July 1988.

Vaughan, Roger J.; Pollard, Robert; and Dyer, Barbara. *The Wealth of the States: Policies for a Dynamic Economy*. Washington, D.C.: The Council of State Planning Agencies, 1984.

Walter, Susan, and Choate, Pat. *Thinking Strategically: A Primer for Public Leaders*. Washington, D.C.: The Council of State Planning Agencies, 1984.

Articles

"_____ Annual Report(s) on the States." *Inc. Magazine* (1979-1987).

Bartik, Timothy J. "Business Location Decisions in the United States: Estimates of the Effects of Unionization, Taxes, and Other Characteristics of States." *Journal of Business & Economic Statistics* (January 1985): 14-22.

Beaumont, Enid F., and Hovey, Harold A. "State, Local, and Federal Economic Development Policies: New Federal Patterns, Chaos, or What?" *Public Administration Review* (March/April 1985): 327-332.

Blair, John P., and Premus, Robert. "Major Factors in Industrial Location: A Review." *Economic Development Quarterly* (February 1987): 72-85.

Bryson, John M., and Roering, William D. "Applying Private Sector Strategic Planning in the Public Sector." *American Planning Association Journal* (Winter 1987): 9-21.

Cetron, Marvin J.; Rocha, Wanda; and Luckins, Rebecca. "Into the 21st Century: Long Term Trends Affecting the United States." *The Futurist* (July/August 1988): 29-40.

Carroll, John J.; Hyde, Mark S.; and Hudson, William E. "State Level Perspectives on Industrial Policy: The Views of Legislators and Bureaucrats." *Economic Development Quarterly* 1 (November 1987): 333-340.

Chi, Keon S. "Strategic Planning for Economic Development: The Wisconsin Experience." *Innovations* (July 1986): 1-8.

Elder, Ann H., and Lind, Nancy S. "The Implications of Uncertainty in Economic Development." *Economic Development Quarterly* (February 1987): 30-40.

Erickson, Rodney A. "Business Climate Studies: A Critical Evaluation." *Economic Development Quarterly* 1 (February 1987) : 62-71, No. 1.

Grady, Dennis O. "State Economic Development Incentives: Why Do States Compete?" *State and Local Government Review* (Fall 1987): 86-94.

Kaufman, Jerome L., and Jacobs, Harvey M. "A Public Planning Perspective on Strategic Planning." *American Planning Association Journal* (Winter 1987): 23-57.

Modic, Stanley J. "Strategic Alliances: A Global Economy Demands Global Partnerships." *Industry Week* (October 3, 1988): 46-49.

Nutt, Paul C., and Backoff, Robert W. "A Strategic Management Process for Public and Third Sector Organizations." *American Planning Association Journal* (Winter 1987): 44-57.

Reed, Christine M.; Reed, B.J.; and Luke, Jeffrey S. "Assessing Readiness for Economic Development Strategic Planning." *American Planning Association Journal* (Autumn 1987): 521-529.

Rubin, Barry M., and Zorn, C. Kurt. "Sensible State and Local Economic Development." *Public Administration Review* (March/April 1985): 333-339.

Sternberg, Ernest. "A Practitioner's Classification of Economic Policy Instruments, With Some Inspiration from Political Economy." *Economic Development Quarterly* 1 (May 1987): 149-161.

Walker, Lee. "Strategic Planning for Economic Development." *State Government News* (November 1988): 16-18.

Warner, Paul D. "Business Climate, Taxes, and Economic Development." *Economic Development Quarterly* 1 (November 1987): 383-390.

Wechsler, Barton, and Backoff, Robert W. "Policy Making and Administration in State Agencies: Strategic Management Approaches." *Public Administration Review* (July/August 1986): 321-322.

Weinstein, Bernard L., and Gross, Harold T. "What Counts Most in the Race for Development." *State Legislatures* (May/June 1988): 22-24.

Wishard, William Van Dusen. "The 21st Century Economy." *The Futurist* (May/June 1987): 23-28.

State Publications

California Department of Commerce. *Facts: The Californias.* Sacramento: 1986.

California Department of Commerce. Office of Business Development. *The Californias: Business Incentives.* Sacramento: February 1988.

California Department of Commerce. Office of Economic Research. *The Californias: Environmental Permits.* Sacramento: 1988.

California Economic Development Corporation. *1987 Annual Report.* Sacramento: 1987.

Florida Department of Commerce. Division of Economic Development. *Project Cornerstone.* Tallahassee: October 1988 (revised).

Hawaii Department of Business and Economic Development. *Envision Hawaii: Strategic Plan: Business and Industry Development, Marketing and Promotion.* Honolulu: January 1988.

Illinois Department of Commerce and Community Affairs. *Sell Illinois: A Strategy for the Present, a Commitment to the Future.* Springfield: 1987.

Illinois Department of Commerce and Community Affairs. *State of Illinois Five Year Economic Development Strategy.* Springfield: March 1988.

Indiana Economic Development Council, Inc. *Looking Back: The Update of Indiana's Economic Development Plan - An Evaluation of Progress to Date, In Step With the Future.* Volume I. Indianapolis: September 1987.

Indiana Economic Development Council, Inc. *Looking Forward: The Update of Indiana's Strategic Economic Development Plan - Strategies for the Future.* Volume II. Indianapolis: June 1988.

Indiana State Chamber of Commerce. *In Step With the Future - Indiana's Strategic Economic Development Plan.* Indianapolis: 1983.

Iowa Department of Economic Development. "Directions for Iowa's Economic Future: Executive Summary." *New Opportunities for Iowa: Strategic Planning Recommendations for Economic Development.* Des Moines: March 1987.

Kentucky Economic Development Cabinet. *Reported Foreign Direct Industrial Investment in Kentucky.* Frankfort: April 1989.

Maine Economic Development Strategy Task Force, Maine Development Foundation. *Establishing the Maine Advantage: An Economic Development Strategy for the State of Maine.* Augusta: October 1987.

Minnesota. *The Report of the Governor's Commission on the Economic Future of Minnesota.* Saint Paul: 1987.

North Carolina Department of Commerce. *North Carolina: The Better Business Climate: Economic Advantages of Warehousing and Distribution.* Raleigh: 1987.

Ohio Department of Development. *Ohio's Thomas Edison Program.* Columbus: August 1985.

Ohio Department of Development. *Resource Ohio: Financial Assistance, Employment and Training, Technical Assistance, Applied Technology and Research.* Columbus, Ohio.

Pennsylvania Economic Development Partnership Office. "Investment in Pennsylvania's Future: The Keystone for Economic Growth." *An Executive Summary of Pennsylvania's Economic Development Partnership Office.* Harrisburg: January 1988.

Rhode Island Planning Program. *Economic Development Strategy.* Providence: March 1986.

South Dakota Governor's Office of Economic Development. *Statewide Action Program for Economic Development.* Pierre: February 1988.

SRI International Center for Economic Competitiveness. *New Seeds for Nebraska: Strategies for Building the Next Economy.* 1987.

SRI International Center for Economic Competitiveness. *Profiles of Key Industries in Pennsylvania: Competing in the Global Marketplace.* January 1988.

Texas Strategic Economic Development Commission. *A Blueprint for Tommorrow's Texas.* Austin: May 1988.

Washington, Office of the Governor. *Washington Economic Development Agenda: Priorities and Strategies.* Olympia: January 1988.

Wisconsin Department of Development. *Final Report: Governor's Advisory Committee on Business Incentives.* Madison: September 1987.

Wisconsin Strategic Development Commission. *Phase I - The Mark of Progress.* Madison: 1985.

Wisconsin Strategic Development Commission. *Wisconsin Strategic Development Commission: The Final Report.* Madison: 1986.

Surveys

The Council of State Governments. "Business Tax and Financial Incentives: A Survey of State Economic Development Agencies (Fifty-State Survey)." Summer and Fall 1988.

The Council of State Governments. "Business Tax and Financial Incentives: A Survey of CSG Corporate Associates." Summer and Fall 1988.